INTRODUCTION TO STATISTICAL MODELLING

INTRODUCTION TO STATISTICAL MODELLING

Annette J. Dobson
UNIVERSITY OF NEWCASTLE,
NEW SOUTH WALES, AUSTRALIA

LONDON NEW YORK
CHAPMAN AND HALL

First published 1983 *by*
Chapman and Hall Ltd
11 *New Fetter Lane, London EC4P 4EE*
Published in the USA by
Chapman and Hall
733 Third Avenue, New York NY10017

Printed in Great Britain by the
University Press, Cambridge

ISBN 0 412 24850 6 (cased)
ISBN 0 412 24860 3 (Science Paperback)

British Library Cataloguing in Publication Data

Dobson, Annette J.
Introduction to statistical modelling.
1. Mathematical statistics 2. Mathematical models
I. Title
519.5 QA276

ISBN 0-412-24850-6
ISBN 0-412-24860-3 Pbk

Library of Congress Cataloging in Publication Data

Dobson, Annette, J., 1945–
Introduction to statistical modelling.

Bibliography: p.
Includes index.
1. Linear models (Statistics) I. Title.
QA276.D59 1983 519.5 83-7495
ISBN 0-412-24850-6
ISBN 0-412-24860-3 (Science paperbacks: pbk.)

CONTENTS

PREFACE

This book is about generalized linear models as described by Nelder and Wedderburn (1972). This approach provides a unified theoretical and computational framework for the most commonly used statistical methods: regression, analysis of variance and covariance, logistic regression, log-linear models for contingency tables and several more specialized techniques. More advanced expositions of the subject are given by McCullagh and Nelder (1983) and Andersen (1980).

The emphasis is on the use of statistical models to investigate substantive questions rather than to produce mathematical descriptions of the data. Therefore parameter estimation and hypothesis testing are stressed.

I have assumed that the reader is familiar with the most commonly used statistical concepts and methods and has some basic knowledge of calculus and matrix algebra. Short numerical examples are used to illustrate the main points.

In writing this book I have been helped greatly by the comments and criticism of my students and colleagues, especially Anne Young. However, the choice of material, and the obscurities and errors are my responsibility and I apologize to the reader for any irritation caused by them.

For typing the manuscript under difficult conditions I am grateful to Anne McKim, Jan Garnsey, Cath Claydon and Julie Latimer.

University of Newcastle,
Australia
December 1982

Annette J. Dobson

1
INTRODUCTION

1.1 Background

This book is designed to introduce the reader to the ideas of statistical modelling and to use these ideas to demonstrate the unity among many commonly used statistical techniques. It is assumed that the reader already has some familiarity with statistical concepts and methods, in particular sampling distributions, hypothesis testing, the use of t-tests, analysis of variance, simple linear regression and chi-squared tests of independence for two-dimensional contingency tables. In addition, some background knowledge of matrix algebra and calculus is assumed.

The reader will find it useful to have access to statistical computing facilities, for example, computer packages such as BMDP (Biomedical Computer Programs, University of California, Los Angeles), MINITAB (Minitab Project, Pennsylvania State University), SAS (SAS Institute, Rayleigh, North Carolina), SPSS (SPSS Inc, Chicago, Illinois), GENSTAT (Rothamsted Experimental Station, Herts, UK) or GLIM (Numerical Algorithms Group, Oxford, UK).

1.2 Scope

The statistical methods considered in this book all involve the analysis of relationships between multiple measurements made on groups of subjects or objects. For example, the measurements might be the heights or weights and ages of boys and girls, or yield and various growing conditions for plants. We use the terms *response*, *outcome* or *dependent variable* for measurements we regard as random variables. These measures are free to vary in response to other variables called *independent*, *explanatory* or *predictor* variables which are non-random measurements or observations (e.g. those fixed by experimental design).

Measurements are made on one of the following scales.

(i) *Nominal* classifications: e.g. red, green, blue; yes, no, do not know, not applicable. In particular, for *binary* or *dichotomous* variables there are only two categories: male, female; dead, alive; smooth leaves, serrated leaves.

(ii) *Ordinal* classifications in which there is some natural order or ranking between the categories: e.g. young, middle aged, old; diastolic blood pressures grouped as $\leqslant 70$, 71–90, 91–110, 111–130, $\geqslant 131$ mm Hg.

(iii) *Continuous* measurements where observations may, at least in theory, fall anywhere on a continuum: e.g. weight, length or time. (This scale includes both *interval* and *ratio scale* measurements – the latter have a well-defined zero.)

Usually nominal and ordinal data are recorded as the numbers of observations in each category. These counts or *frequencies* are called *discrete* variables. For continuous data the individual measurements are recorded. The term *quantitative* is often used for a variable measured on a continuous scale and the term *qualitative* for nominal and sometimes for ordinal measurements. A qualitative, explanatory variable is called a *factor* and its categories are called the *levels* of the factor. A quantitative explanatory variable is called a *covariate*.

Methods of statistical analysis depend on the measurement scales of the response and explanatory variables. In practice ordinal data, because of their intermediate position between nominal and continuous observations, are often analysed by methods designed for one of the other two scales. In this book methods specific to ordinal measurements are rarely considered due to limitations of space rather than as an endorsement of methods which are not strictly appropriate.

Generally we consider only those statistical methods which are relevant when there is just *one response variable* although there will usually be several explanatory variables. For various combinations of response and explanatory variables Table 1.1 shows the main methods of statistical analysis and the chapters in which these are considered.

Chapters 2 to 5 cover the common theoretical approaches needed in the subsequent chapters which focus on methods for analysing particular kinds of data.

Chapter 2 develops the idea of statistical modelling via some numerical examples. The modelling process involves three steps:

(i) specifying plausible equations and probability distributions (models) to describe the main features of the response variable;

(ii) estimating parameters used in the models;

(iii) making inferences; for example, testing hypotheses by considering how adequately the models fit the actual data.

In the numerical examples the modelling approach is compared with more traditional analyses of the same data sets.

Table 1.1 Major methods of statistical analysis for response and explanatory variables measured on various scales.

<table>
<tr><th rowspan="2">Explanatory variables</th><th colspan="3">Response variable</th></tr>
<tr><th>Binary</th><th>Nominal with more than 2 categories</th><th>Continuous</th></tr>
<tr><td>Binary</td><td rowspan="4">Comparing proportions, dose–response models and logistic regression (Ch. 8)</td><td rowspan="2">Contingency tables and log-linear models (Ch. 9)</td><td rowspan="2">Analysis of variance (Ch. 7)</td></tr>
<tr><td>Nominal with more than 2 categories</td></tr>
<tr><td>Continuous</td><td rowspan="2"></td><td rowspan="2">Multiple regression and analysis of covariance (Ch. 6 and 7)</td></tr>
<tr><td>Mixed</td></tr>
</table>

The next three chapters concern the theoretical development of each of the three steps just outlined. Chapter 3 is about the exponential family of distributions, which includes the Normal, Poisson and binomial distributions. It also covers generalized linear models (as defined by Nelder and Wedderburn, 1972) which include linear regression and many other models as special cases. In Chapter 4 two methods of estimation, maximum likelihood and least squares, are considered. For some generalized linear models these methods give identical results but for others likelihood methods are often more useful. Chapter 5 concerns test statistics which provide measures of how well a model describes a given data set. Hypothesis testing is carried out by specifying alternative models (one corresponding to the null hypothesis and the other to a more general hypothesis) and then comparing test statistics which measure the 'goodness of fit' of each model. Typically the model corresponding to the null hypothesis is simpler, so if it fits the data about as well as the other model we usually prefer it on the grounds of parsimony (i.e. we retain the null hypothesis).

Chapter 6 is about multiple linear regression which is the standard method for relating a continuous response variable to several continuous explanatory (or predictor) variables. Analysis of variance (ANOVA) and analysis of covariance (ANCOVA) are discussed in Chapter 7. ANOVA is used for a continuous response variable and qualitative explanatory variables (factors). ANCOVA is used when at least one of the explanatory variables is qualitative and at least one is continuous. This distinction between multiple linear regression and ANCOVA (or even ANOVA) is somewhat artificial. The methods are so closely related that nowadays it is common to use the same computational tools for all such problems. The terms multiple regression or *general linear model* are used to cover the spectrum of methods for analysing one continuous response variable and multiple explanatory variables.

Chapter 8 is about methods for analysing binary response data. The most common one is logistic regression which is used to model relationships between the response variable and several explanatory variables which may be categorical or continuous. Methods for relating the response to a single continuous variable, the dose, are also considered; these include probit analysis which was originally developed for analysing dose–response data from bioassays.

Chapter 9 concerns contingency tables and is mainly about log-linear models which are used to investigate relationships between several categorical variables. In this chapter the distinction between response and explanatory variables is less crucial and the methods are also suitable for multiple responses.

There are other statistical methods which fit into the same general framework as the major methods considered in this book but they are less widely used so they are not covered here. They include models for survival data, inverse polynomial response surfaces, variance components, the Rasch model for test items and the Bradley–Terry model for paired comparisons. These methods are discussed in the books by McCullagh and Nelder (1983) and Andersen (1980).

1.3 Notation

Generally we follow the convention of denoting random variables by upper case italic letters and observed values by the corresponding lower case letters; for example, the observations $y_1, y_2, \ldots, y_N$ are regarded as realizations of the random variables $Y_1, Y_2, \ldots, Y_N$. Greek letters are used to denote parameters and the corresponding lower case roman letters to denote estimators and estimates; occasionally the symbol $\hat{}$ is used for estimators or estimates. For example, the parameter β is estimated by $\hat{\beta}$ or b. Sometimes these conventions are not strictly adhered to, either to avoid excessive notation in cases when the meaning should be apparent from the context, or when there is a strong tradition of alternative notation (e.g. e or ϵ for random error terms).

Vectors and matrices, whether random or not, are denoted by bold face lower and upper case letters, respectively; for example, **y** represents a vector of observations

$$\begin{bmatrix} y_1 \\ \vdots \\ y_N \end{bmatrix}$$

or a vector of random variables

$$\begin{bmatrix} Y_1 \\ \vdots \\ Y_N \end{bmatrix}$$

$\boldsymbol{\beta}$ denotes a vector of parameters, and $\mathbf{X}$ is a matrix. The superscript ^{T} is used for matrix transpose or when a column vector is written as a row, e.g. $\mathbf{y} = [y_1, \ldots, y_N]^{\mathrm{T}}$.

The probability density function of a continuous random variable Y (or the probability distribution if Y is discrete) is denoted by

$$f(y; \boldsymbol{\theta})$$

where $\boldsymbol{\theta}$ represents the parameters of the distribution.

We use dot (·) subscripts for summation and bars (—) for means, thus

$$\bar{y} = \frac{1}{N} \sum_{i=1}^{N} y_i = \frac{1}{N} y_{\cdot}.$$

1.4 Distributions derived from the Normal distribution

The sampling distributions of many of the statistics used in this book depend on the Normal distribution. They do so either directly, because they are based on Normally distributed random variables, or asymptotically, via the Central Limit Theorem for large samples. In this section we give definitions and notation for these distributions and summarize the relationships between them.

If the random variable Y has the *Normal distribution* with mean μ and variance σ^2 we denote this by $Y \sim \mathrm{N}(\mu, \sigma^2)$; $\mathrm{N}(0, 1)$ is called the *standard Normal distribution.* For *independent* random variables $Y_1, \ldots, Y_n$, if $Y_i \sim \mathrm{N}(\mu_i, \sigma_i^2)$ for $i = 1, \ldots, n$ we sometimes write this as $Y_i \sim \mathrm{NID}(\mu_i, \sigma_i^2)$. More generally, if $Y_i \sim \mathrm{N}(\mu, \sigma_i^2)$ for $i = 1, \ldots, n$ and the covariance of Y_i and Y_j is $\mathrm{cov}(Y_i, Y_j) = \sigma_{ij}^2$, the joint distribution of the Ys is the *multivariate Normal distribution* with mean $\boldsymbol{\mu} = [\mu_1, \ldots, \mu_n]^{\mathrm{T}}$ and variance–covariance matrix $\mathbf{V}$ with elements σ_{ij}^2; we write this as

$$\mathbf{y} \sim \mathrm{N}(\boldsymbol{\mu}, \mathbf{V})$$

where $\mathbf{y} = [Y_1, \ldots, Y_n]^{\mathrm{T}}$. Any linear combination of Normally distributed random variables is also Normally distributed, for example, if $Y_i \sim \mathrm{NID}(\mu_i, \sigma_i^2)$ then

$$W = \sum_{i=1}^{n} a_i Y_i \sim \mathrm{N}\left(\sum_{i=1}^{n} a_i \mu_i, \sum_{i=1}^{n} a_i^2 \sigma_i^2 \right)$$

where the a_is are constants.

The *central chi-squared distribution* with *n degrees of freedom* is defined as the sum of squares of n independent random variables $Z_1, \ldots, Z_n$ each with the standard Normal distribution. It is denoted by

$$X^2 = \sum_{i=1}^{n} Z_i^2 \sim \chi_n^2.$$

In matrix notation this is $X^2 = \mathbf{z}^T\mathbf{z} \sim \chi_n^2$. If $Y_1, \ldots, Y_n$ are independent Normally distributed random variables all with the same variance, i.e. $Y_i \sim \text{NID}(\mu_i, \sigma^2)$, then

$$X^2 = \sum_{i=1}^{n} \left(\frac{Y_i - \mu_i}{\sigma}\right)^2 \sim \chi_n^2 \tag{1.1}$$

because $Z_i = (Y_i - \mu_i)/\sigma \sim \text{N}(0, 1)$. More generally if $\mathbf{y} \sim \text{N}(\boldsymbol{\mu}, \mathbf{V})$ where $\mathbf{V}$ is non-singular with inverse $\mathbf{V}^{-1}$, then

$$X^2 = (\mathbf{y} - \boldsymbol{\mu})^T \mathbf{V}^{-1} (\mathbf{y} - \boldsymbol{\mu}) \sim \chi_n^2. \tag{1.2}$$

If $\mathbf{y} \sim \text{N}(\boldsymbol{\mu}, \mathbf{V})$ then the distribution of $\mathbf{y}^T\mathbf{V}^{-1}\mathbf{y}$ is called the *non-central chi-squared distribution* with n degrees of freedom and non-centrality parameter $\lambda = \frac{1}{2}\boldsymbol{\mu}^T\mathbf{V}^{-1}\boldsymbol{\mu}$. We denote this by $\mathbf{y}^T\mathbf{V}^{-1}\mathbf{y} \sim \chi^2(n, \lambda)$. If $X_1^2, \ldots, X_m^2$ are independent random variables with $X_i^2 \sim \chi^2(n_i, \lambda_i)$ then

$$\sum_{i=1}^{m} X_i^2 \sim \chi^2\left(\sum_{i=1}^{m} n_i, \sum_{i=1}^{m} \lambda_i\right).$$

This is the *reproductive property* of the chi-squared distribution. If $\mathbf{y} \sim \text{N}(\boldsymbol{\mu}, \mathbf{V})$ where $\mathbf{y}$ has n elements and $\mathbf{V}$ is singular with rank $k < n$, then $\mathbf{y}^T\mathbf{V}^-\mathbf{y}$ has the chi-squared distribution with k degrees of freedom and non-centrality parameter $\frac{1}{2}\boldsymbol{\mu}^T\mathbf{V}^-\boldsymbol{\mu}$, where $\mathbf{V}^-$ denotes a generalized inverse of $\mathbf{V}$.

The *t-distribution* with n degrees of freedom is defined by the ratio of random variables

$$T = \frac{Z}{(X^2/n)^{\frac{1}{2}}} \tag{1.3}$$

in which $Z \sim \text{N}(0, 1)$, $X^2 \sim \chi_n^2$ and Z and X^2 are independent. It is denoted by $T \sim t_n$.

The *central F-distribution* with n and m degrees of freedom is defined by the ratio

$$F = \frac{X_1^2}{n} \bigg/ \frac{X_2^2}{m}, \tag{1.4}$$

where X_1^2 and X_2^2 are independent random variables with $X_1^2 \sim \chi_n^2$ and $X_2^2 \sim \chi_m^2$; it is denoted by $F \sim F_{n,m}$. Hence the relationship between the t-distribution and the F-distribution obtained from (1.3), is

$$T^2 = \frac{Z^2}{1} \bigg/ \frac{X^2}{n} \sim F_{1,n}. \tag{1.5}$$

The *non-central F-distribution* is defined as the ratio of two independent random variables, each divided by its degrees of freedom, where the numerator has a non-central chi-squared distribution and the denominator has a central chi-squared distribution, i.e.

$$F = \frac{X_1^2}{n} \bigg/ \frac{X_2^2}{m}$$

where $X_1^2 \sim \chi^2\,(n, \tfrac{1}{2}\boldsymbol{\mu}^{\mathrm{T}}\mathbf{V}^{-1}\boldsymbol{\mu})$ and $X_2^2 \sim \chi_m^2$.

2
MODEL FITTING

2.1 Introduction

The transmission and reception of information involves a message, or *signal*, which is distorted by *noise*. It is sometimes useful to think of scientific data as measurements composed of signal and noise and to construct mathematical models incorporating both of these components. Often the signal is regarded as *deterministic* (i.e. non-random) and the noise as random. Therefore, a mathematical model of the data combining both signal and noise is probabilistic and it is called a statistical model.

Another way of thinking of a statistical model is to consider the signal component as a mathematical description of the main features of the data and the noise component as all those characteristics not 'explained' by the model (i.e. by its signal component).

Our goal is to extract from the data as much information as possible about the signal as it is defined by the model. Typically the mathematical description of the signal involves several unknown constants, termed *parameters*. The first step is to estimate values for these parameters from the data.

Once the signal component has been quantified we can partition the total variability observed in the data into a portion attributable to the signal and the remainder attributable to the noise. A criterion for a good model is one which 'explains' a large proportion of this variability, i.e. one in which the part attributable to signal is large relative to the part attributable to noise. In practice, this has to be balanced against other criteria such as simplicity. Occam's Razor suggests that a parsimonious model which describes the data adequately may be preferable to a complicated one which leaves little of the variability 'unexplained'.

In many situations we wish to test hypotheses about the parameters. This can be performed in the context of model fitting by defining a series of different models corresponding to different hypotheses. Then the question about whether the data support a particular hypothesis can be formulated in terms of the adequacy of fit of the corresponding model (i.e. the amount of variability it explains) relative to other models.

These ideas are now illustrated by two detailed examples.

2.2 Plant growth example

Suppose that genetically similar seeds are randomly assigned to be raised either in a nutritionally enriched environment (treatment) or under standard conditions (control) using a *completely randomized experimental design.* After a predetermined period all plants are harvested, dried and weighed. The results, expressed as dried weight in grams, for samples of 10 plants from each environment are given in Table 2.1. Fig. 2.1 shows the distributions of these weights.

Table 2.1 Plant weights from two different growing conditions.

Control (1)	4.17	5.58	5.18	6.11	4.50	4.61	5.17	4.53	5.33	5.14
Treatment (2)	4.81	4.17	4.41	3.59	5.87	3.83	6.03	4.89	4.32	4.69

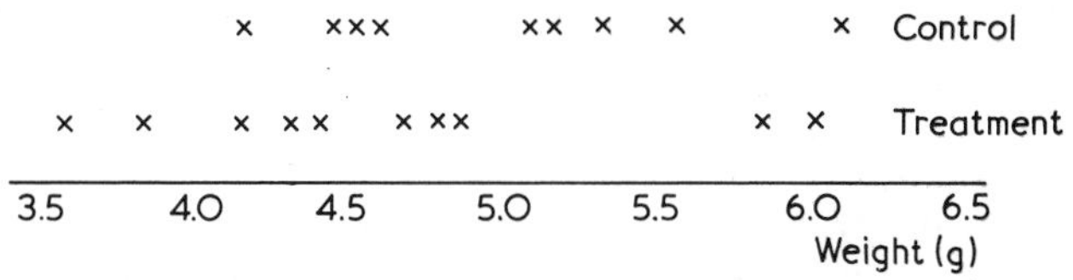

Figure 2.1 Plant growth data from Table 2.1.

The first step is to formulate models to describe these data, for example

$$Y_{jk} = \mu_j + e_{jk}, \tag{2.1}$$

where

(i) Y_{jk} is the weight of the kth plant ($k = 1, \ldots, K$ with $K = 10$ in this case) from the jth sample ($j = 1$ for control and $j = 2$ for treatment);

(ii) μ_j is a parameter, the signal component of weight, determined by the growth environment. It represents a common characteristic of all plants grown under the conditions experienced by sample j;

(iii) e_{jk} is the noise component. It is a random variable (although by convention it is usually written using the lower case). It is sometimes called the *random error term.* It represents that element of weight unique to the kth observation from sample j.

From the design of the experiment we assume that the e_{jk}s are independent and identically distributed with the Normal distribution with mean zero and variance σ^2, i.e. $e_{jk} \sim \text{NID}(0, \sigma^2)$ and therefore $Y_{jk} \sim \text{NID}(\mu_j, \sigma^2)$ for all j and k.

We would like to know if the enriched environment made a difference to the weight of the plants so we need to estimate the difference between μ_1 and μ_2 and test whether it differs significantly from some pre-specified value (such as zero).

An alternative specification of the model which is more suitable for comparative use is

$$Y_{jk} = \mu + \alpha_j + e_{jk} \tag{2.2}$$

where

(i) Y_{jk} and e_{jk} are defined as before;
(ii) μ is a parameter representing that aspect of growth common to both environments; and
(iii) α_1 and α_2 are parameters representing the differential effects due to the control or treatment conditions; formally $\alpha_j = \mu_j - \mu$ for $j = 1, 2$.

If the nutritionally enriched conditions do not enhance (or inhibit) plant growth, then the terms α_j will be negligible and so the model (2.2) will be equivalent to

$$Y_{jk} = \mu + e_{jk}. \tag{2.3}$$

Therefore, testing the hypothesis that there is no difference in weight due to the different environments (i.e. $\mu_1 = \mu_2$ or equivalently $\alpha_1 = \alpha_2 = 0$) is equivalent to comparing the adequacy of (2.1) and (2.3) for describing the data.

The next step is to estimate the model parameters. We will do this using the *likelihood function* which is the same as the joint probability density function of the response variables Y_{jk} but viewed primarily as a function of the parameters, conditional on the observations. *Maximum likelihood estimators* are the estimators which correspond to the maximum value of the likelihood function or, equivalently, its logarithm which is called the *log-likelihood function.*

We begin by estimating parameters μ_1 and μ_2 in (2.1) treating σ^2 as a known constant (in this context σ^2 is often referred to as a *nuisance parameter*). Since the Y_{jk}s are independent, the likelihood function is

$$\prod_{j=1}^{2} \prod_{k=1}^{K} \frac{1}{(2\pi\sigma^2)^{\frac{1}{2}}} \exp\left\{-\frac{1}{2\sigma^2}(y_{jk} - \mu_j)^2\right\}$$

and the log-likelihood function is

$$l_1 = -K \log(2\pi\sigma^2) - \frac{1}{2\sigma^2} \sum_{j=1}^{2} \sum_{k=1}^{K} (y_{jk} - \mu_j)^2,$$

so the maximum likelihood estimators of μ_1 and μ_2 are given by the solutions of

$$\frac{\partial l_1}{\partial \mu_j} = \frac{1}{\sigma^2} \sum_{k=1}^{K} (y_{jk} - \mu_j) = 0, \quad j = 1, 2$$

i.e.

$$\hat{\mu}_j = \frac{1}{K} \sum_{k=1}^{K} y_{jk} = \frac{1}{K} y_{j\cdot} = \bar{y}_j \quad \text{for } j = 1, 2.$$

By considering the second derivatives it can be verified that $\hat{\mu}_1$ and $\hat{\mu}_2$ correspond to the maximum of l_1. Let

$$\hat{l}_1 = -K\log(2\pi\sigma^2) - \frac{1}{2\sigma^2}\hat{S}_1$$

denote the maximum of l_1 where $\hat{S}_1 = \sum_{j=1}^{2}\sum_{k=1}^{K}(y_{jk}-\bar{y}_j)^2$.

For the model given by (2.3) the likelihood function is

$$\prod_{j=1}^{2}\prod_{k=1}^{K}\frac{1}{(2\pi\sigma^2)^{\frac{1}{2}}}\exp\left\{-\frac{1}{2\sigma^2}(y_{jk}-\mu)^2\right\}$$

since $Y_{jk} \sim \text{NID}(\mu, \sigma^2)$ for $j = 1, 2$ and $k = 1, \ldots, K$. Therefore the log-likelihood function is

$$l_0 = -K\log(2\pi\sigma^2) - \frac{1}{2\sigma^2}\sum_{j=1}^{2}\sum_{k=1}^{K}(y_{jk}-\mu)^2,$$

and so the estimator $\hat{\mu}$ obtained from the solution of $\partial l_0/\partial\mu = 0$ is

$$\hat{\mu} = \frac{1}{2K}\sum_{j=1}^{2}\sum_{k=1}^{K}y_{jk} = \frac{1}{2K}y_{..} = \bar{y} = \tfrac{1}{2}(\bar{y}_1+\bar{y}_2).$$

Hence the maximum of l_0 is

$$\hat{l}_0 = -K\log(2\pi\sigma^2) - \frac{1}{2\sigma^2}\hat{S}_0$$

where

$$\hat{S}_0 = \sum_{j=1}^{2}\sum_{k=1}^{K}(y_{jk}-\bar{y})^2.$$

For the plant data the values of the maximum likelihood estimates and the statistics $\hat{S}_1$ and $\hat{S}_0$ are shown in Table 2.2.

Table 2.2 Analysis of plant growth data in Table 2.1.

Model (2.1):	$\hat{\mu}_1 = 5.032, \hat{\mu}_2 = 4.661$ and $\hat{S}_1 = 8.729$
Model (2.3):	$\hat{\mu} = 4.8465$ and $\hat{S}_0 = 9.417$

The third step in the model fitting procedure involves testing hypotheses. If the null hypothesis $H_0: \mu_1 = \mu_2$ is correct then the models (2.1) and (2.3) are the same so the maximum values $\hat{l}_1$ and $\hat{l}_0$ of the log-likelihood functions should be nearly equal, or equivalently, $\hat{S}_1$ and $\hat{S}_0$ should be nearly equal. If the data support this hypothesis, we would feel justified in using the simpler model (2.3) to describe the data. On the other hand, if the more general hypothesis $H_1: \mu_1$ and μ_2 are not necessarily equal, is true then $\hat{S}_0$ should be larger than $\hat{S}_1$ (corresponding to $\hat{l}_0$ smaller than $\hat{l}_1$) and the model given by (2.1) would be preferable.

To assess the relative magnitude of $\hat{S}_1$ and $\hat{S}_0$ we need to consider the sampling distributions of the corresponding random variables

$$S_1 = \sum_{j=1}^{2}\sum_{k=1}^{K}(Y_{jk}-\bar{Y}_j)^2 \quad \text{and} \quad S_0 = \sum_{j=1}^{2}\sum_{k=1}^{K}(Y_{jk}-\bar{Y})^2.$$

It can be shown that

$$\frac{1}{\sigma^2}S_1 = \frac{1}{\sigma^2}\sum_{j=1}^{2}\sum_{k=1}^{K}(Y_{jk}-\bar{Y}_j)^2 = \frac{1}{\sigma^2}\sum_{j=1}^{2}\sum_{k=1}^{K}(Y_{jk}-\mu_j)^2 - \frac{K}{\sigma^2}\sum_{j=1}^{2}(\bar{Y}_j-\mu_j)^2.$$

For the more general model (2.1) we assume that $Y_{jk} \sim \text{NID}(\mu_j, \sigma^2)$ and so $\bar{Y}_j \sim \text{NID}(\mu_j, \sigma^2/K)$. Therefore (S_1/σ^2) is the difference between the sum of the squares of $2K$ independent random variables $(Y_{jk}-\mu_j)/\sigma$ which each has the distribution N(0, 1) and the sum of two independent random variables $(\bar{Y}_j-\mu_j)/(\sigma^2/K)^{\frac{1}{2}}$ which also have the N(0, 1) distribution. Hence, from definition (1.1),

$$\frac{1}{\sigma^2}S_1 \sim \chi^2_{2K-2}.$$

Similarly for the simpler model (2.3), let $\bar{\mu} = \frac{1}{2}(\mu_1+\mu_2)$ then

$$\frac{1}{\sigma^2}S_0 = \frac{1}{\sigma^2}\sum_{j=1}^{2}\sum_{k=1}^{K}(Y_{jk}-\bar{Y})^2$$

$$= \frac{1}{\sigma^2}\sum_{j=1}^{2}\sum_{k=1}^{K}(Y_{jk}-\bar{\mu}) - \frac{2K}{\sigma^2}(\bar{Y}-\bar{\mu})^2.$$

If $Y_{jk} \sim \text{NID}(\mu_j, \sigma^2)$ then $\bar{Y} \sim \text{N}(\bar{\mu}, \sigma^2/2K)$. Also if $\mu_1 = \mu_2 = \bar{\mu}$ (corresponding to H_0) then the first term of (S_0/σ^2) is the sum of the squares of $2K$ independent random variables $(Y_{jk}-\bar{\mu})/\sigma \sim \text{N}(0, 1)$ and therefore

$$\frac{1}{\sigma^2}S_0 \sim \chi^2_{2K-1}.$$

However, if μ_1 and μ_2 are not necessarily equal (corresponding to H_1) then $(Y_{jk}-\bar{\mu})/\sigma \sim \text{N}(\mu_j-\bar{\mu}, 1)$ so that (S_0/σ^2) has a non-central chi-squared distribution with $2K-1$ degrees of freedom.

The statistic S_0-S_1 represents the difference in fit between the two models. If $\text{H}_0\colon \mu_1 = \mu_2$, is correct then

$$\frac{1}{\sigma^2}(S_0-S_1) \sim \chi^2_1;$$

otherwise it has a non-central chi-squared distribution. However, since σ^2 is unknown we cannot compare S_0-S_1 directly with the χ^2_1 distribution. Instead we eliminate σ^2 by using the ratio of $(S_0 - S_1)/\sigma^2$ and the central chi-squared random variable (S_1/σ^2), each divided by its degrees of freedom, i.e.

$$f = \frac{1}{\sigma^2}\frac{(S_0 - S_1)}{1} \Big/ \frac{1}{\sigma^2}\frac{S_1}{2K-2} = \frac{S_0 - S_1}{S_1/(2K-2)}.$$

If H_0 is correct, by definition (1.4), f has the central F-distribution with 1 and $(2K-2)$ degrees of freedom; otherwise f has a non-central F-distribution and so it is likely to be larger than predicted by $F_{1,2K-2}$.

For the plant weight data,

$$f = \frac{9.417 - 8.729}{8.729/18} = 1.42$$

which is not statistically significant when compared with the $F_{1,18}$ distribution. Thus the data provide no evidence against H_0 so we conclude that there is probably no difference in weight due to the different environmental conditions and we can use the simpler model (2.3) to describe the data.

The more conventional approach to testing H_0 against H_1 is to use a t-test, i.e. to calculate

$$T = \frac{\bar{Y}_1 - \bar{Y}_2}{s\left(\dfrac{1}{K} + \dfrac{1}{K}\right)^{\frac{1}{2}}}$$

where s^2, the pooled variance, is

$$s^2 = \frac{1}{2K-2}\sum_{j=1}^{2}\sum_{k=1}^{K}(Y_{jk} - \bar{Y}_j)^2 = \frac{1}{2K-2}S_1.$$

If H_0 is correct the statistic T has the distribution t_{2K-2}. The relationship between the test statistics T and f is obtained as follows:

$$T^2 = \frac{(\bar{Y}_1 - \bar{Y}_2)^2}{2s^2/K} = \frac{K(\bar{Y}_1 - \bar{Y}_2)^2}{2\,S_1/(2K-2)},$$

but

$$S_0 - S_1 = \sum_{j=1}^{2}\sum_{k=1}^{K}[(Y_{jk} - \bar{Y})^2 - (Y_{jk} - \bar{Y}_j)^2]$$

$$= \tfrac{1}{2}K(\bar{Y}_1 - \bar{Y}_2)^2$$

so that

$$T^2 = \frac{S_0 - S_1}{S_1/(2K-2)} = f$$

corresponding to the distributional relationship that if $T \sim t_n$ then $T^2 \sim F_{1,n}$ (see (1.5)).

The advantages of using an F-test instead of a t-test are:

(i) it can be generalized to test the equality of more than two means;

(ii) it is more closely related to the general methods considered in this book which involve comparing statistics that measure the 'goodness of fit' of competing models.

2.3 Birthweight example

The data in Table 2.3 are the birthweights (g) and estimated gestational ages (weeks) of 12 male and female babies born in a certain hospital. The mean ages are almost the same for both sexes but the mean birthweight for males is higher than for females. The data are plotted in Fig. 2.2; they suggest a linear trend of birthweight increasing with gestational age. The question of interest is whether the rate of increase is the same for males and females.

Table 2.3 Birthweight and gestational age for male and female babies

	Male		Female	
	Age (weeks)	Birthweight (g)	Age (weeks)	Birthweight (g)
	40	2968	40	3317
	38	2795	36	2729
	40	3163	40	2935
	35	2925	38	2754
	36	2625	42	3210
	37	2847	39	2817
	41	3292	40	3126
	40	3473	37	2539
	37	2628	36	2412
	38	3176	38	2991
	40	3421	39	2875
	38	2975	40	3231
Means	38.33	3024.00	38.75	2911.33

A fairly general statistical model for these data is

$$Y_{jk} = \alpha_j + \beta_j x_{jk} + e_{jk}, \qquad (2.4)$$

where

(i) the response Y_{jk} is the birthweight for the kth baby of sex j where $j = 1$ for males, $j = 2$ for females and $k = 1, \ldots, K = 12$;
(ii) the parameters α_1 and α_2 represent the intercepts of the lines for the two sexes;
(iii) the parameters β_1 and β_2 represent the slopes or rates of increase for the two sexes;
(iv) the independent variable x_{jk} is the age of the (j, k)th baby (it is not a random variable);

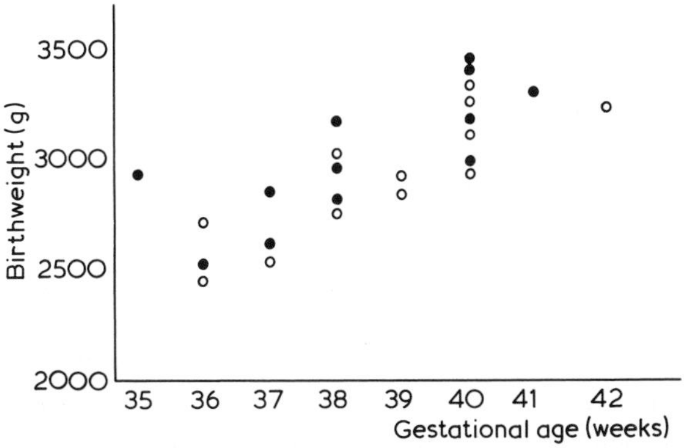

Figure 2.2 Birthweight and gestational age for male and female babies. ○, male; ●, female.

(v) the random error term is e_{jk}; we assume that $e_{jk} \sim \text{NID}(0, \sigma^2)$ for all j and k.

If the rate of increase is the same for males and females then the simpler model

$$Y_{jk} = \alpha_j + \beta x_{jk} + e_{jk} \tag{2.5}$$

is appropriate, where the one parameter β in (2.5) corresponds to the two parameters β_1 and β_2 in (2.4). Thus we can test the null hypothesis

$$\text{H}_0 : \beta_1 = \beta_2 \; (= \beta)$$

against the more general hypothesis

$$\text{H}_1 : \beta_1 \text{ and } \beta_2 \text{ not necessarily equal,}$$

by comparing how well the models (2.4) and (2.5) fit the data.

The next step in the modelling process is to estimate the parameters. For this example we will use the *method of least squares* instead of the method of maximum likelihood. It consists of minimizing the sum of squares of the differences between the responses and their expected values. For the model (2.4) $E(Y_{jk}) = \alpha_j + \beta_j x_{jk}$ because we assumed that $E(e_{jk}) = 0$ so

$$S = \sum_j \sum_k (Y_{jk} - \alpha_j - \beta_j x_{jk})^2.$$

Geometrically, S is the sum of squares of the vertical distances from the points (x_{jk}, y_{jk}) to the line $y = \alpha_j + \beta_j x$ (Fig. 2.3). Algebraically it is the sum of squares of the error terms,

$$S = \sum_j \sum_k e_{jk}^2.$$

Estimators derived by minimizing S are called *least squares estimators* and the minimum value of S is a measure of the fit of the model. An advantage

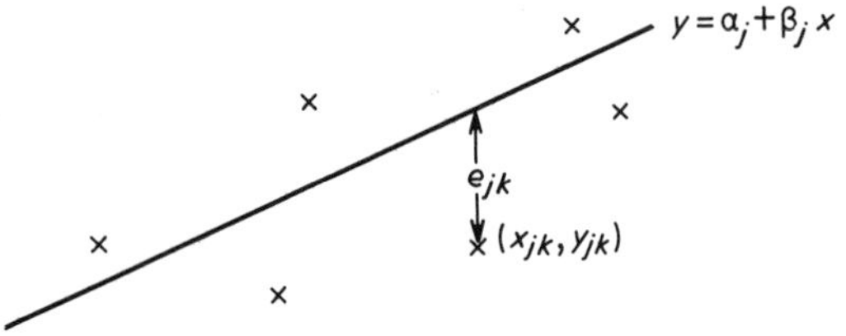

Figure 2.3 Distance from a point (x_{jk}, y_{jk}) to the line $y = \alpha_j + \beta_j x$.

of this method is that it does not require detailed assumptions about the distribution of the error terms (for example, that they are Normally distributed). However, such assumptions are required in order to compare minimum values of S obtained from different models.

Firstly, for (2.4)

$$S_1 = \sum_{j=1}^{2} \sum_{k=1}^{K} (Y_{jk} - \alpha_j - \beta_j x_{jk})^2,$$

so the least squares estimators for the parameters are the solutions of

$$\frac{\partial S_1}{\partial \alpha_j} = -2 \sum_{k=1}^{K} (Y_{jk} - \alpha_j - \beta_j x_{jk}) = 0,$$

$$\frac{\partial S_1}{\partial \beta_j} = -2 \sum_{k=1}^{K} x_{jk}(Y_{jk} - \alpha_j - \beta_j x_{jk}) = 0, \quad \text{for } j = 1, 2.$$

These equations can be simplified to the form

$$\left.\begin{aligned} \sum_{k=1}^{K} Y_{jk} - K\alpha_j - \beta_j \sum_{k=1}^{K} x_{jk} &= 0 \\ \sum_{k=1}^{K} x_{jk} Y_{jk} - \alpha_j \sum_{k=1}^{K} x_{jk} - \beta_j \sum_{k=1}^{K} x_{jk}^2 &= 0 \end{aligned}\right\} j = 1, 2$$

when they are called the *normal equations*. The solutions are

$$b_j = \frac{K \sum_k x_{jk} y_{jk} - \left(\sum_k x_{jk}\right)\left(\sum_k y_{jk}\right)}{K \sum_k x_{jk}^2 - \left(\sum_k x_{jk}\right)^2}$$

$$a_j = \bar{y}_j - b_j \bar{x}_j$$

for $j = 1, 2$. Then the value for S_1 is

$$\hat{S}_1 = \sum_{j=1}^{2} \sum_{k=1}^{K} (y_{jk} - a_j - b_j x_{jk})^2.$$

Secondly, for (2.5)

$$S_0 = \sum_{j=1}^{2} \sum_{k=1}^{K} (Y_{jk} - \alpha_j - \beta x_{jk})^2,$$

so the least squares estimators are the solutions of

$$\frac{\partial S_0}{\partial \alpha_j} = -2 \sum_k (Y_{jk} - \alpha_j - \beta x_{jk}) = 0, \qquad j = 1, 2,$$

and

$$\frac{\partial S_0}{\partial \beta} = -2 \sum_j \sum_k x_{jk} (Y_{jk} - \alpha_j - \beta x_{jk}) = 0.$$

Hence

$$b = \frac{K \sum_j \sum_k x_{jk} y_{jk} - \sum_j \left(\sum_k x_{jk} \sum_k y_{jk} \right)}{K \sum_j \sum_k x_{jk}^2 - \sum_j \left(\sum_k x_{jk} \right)^2}$$

and $a_j = \bar{y}_j - b\bar{x}_j$.

For the birthweight example, the data are summarized in Table 2.4 (summation is over $k = 1, \ldots, K$ with $K = 12$). The least squares estimates for both models are given in Table 2.5.

Table 2.4 Summary of birthweight data in Table 2.3.

	Male, $j = 1$	Female, $j = 2$
Σx	460	465
Σy	36288	34936
Σx^2	17672	18055
Σy^2	110623496	102575468
Σxy	1395370	1358497

Table 2.5 Analysis of birthweight data in Table 2.3.

Model (2.4):	$b_1 = 111.983$,	$a_1 = -1268.672$,	
	$b_2 = 130.400$,	$a_2 = -2141.667$,	$\hat{S}_1 = 652424.5$
Model (2.5):	$b = 120.894$,	$a_1 = -1610.283$,	
		$a_2 = -1773.322$,	$\hat{S}_0 = 658770.8$

To test the hypothesis $H_0: \beta_1 = \beta_2$, i.e. to compare the models given by (2.4) and (2.5), we need to know the sampling distribution of the minimum of the sum of squares, S. By analogous arguments to those used in the previous example, it can be shown that $(S_1/\sigma^2) \sim \chi^2_{20}$ and if H_0 is correct then $(S_0/\sigma^2) \sim \chi^2_{21}$. In each case the number of degrees of freedom is the number of observations minus the number of parameters estimated. The improvement in fit for (2.4) compared with (2.5) is

$$\frac{1}{\sigma^2}(S_0 - S_1),$$

which can be compared with fit of the more detailed model (2.4), i.e. with (S_1/σ^2) using the test statistic

$$f = \frac{(S_0 - S_1)/1}{S_1/(2K-4)}.$$

If the hypothesis H_0 is correct, $f \sim F_{1,\,2K-4}$. For these data the value of f is 0.2 which is certainly not statistically significant, so the data provide no evidence against the hypothesis $\beta_1 = \beta_2$ and we have reason for preferring the simpler model given by (2.5).

2.4 Notation for linear models

The models considered in the above examples can be written in matrix notation in the form

$$\mathbf{y} = \mathbf{X}\boldsymbol{\beta} + \mathbf{e} \qquad (2.6)$$

where

(i) $\mathbf{y}$ is a vector of responses,
(ii) $\boldsymbol{\beta}$ is a vector of parameters,
(iii) $\mathbf{X}$ is a matrix whose elements are zeros or ones or values of 'independent' variables, and
(iv) $\mathbf{e}$ is a vector of random error terms.

For quantitative explanatory variables (e.g. age in the birthweight example) the model contains terms of the form βx where the parameter β represents the rate of change in the response corresponding to changes in the independent variable x.

For qualitative explanatory variables there is a parameter to represent each level of a factor (e.g. the effects due to environmental conditions in the plant growth example). The corresponding elements of $\mathbf{X}$ are chosen to exclude or include the appropriate parameters for each observation; they are called *dummy variables* (if only zeros and ones are used for $\mathbf{X}$ the term *indicator variable* is used).

Example 2.1

For the plant growth example the more general model was

$$Y_{jk} = \mu_j + e_{jk}; \qquad j = 1, 2 \text{ and } k = 1, \ldots, K.$$

The corresponding elements of (2.6) are

$$\mathbf{y} = \begin{bmatrix} Y_{11} \\ Y_{12} \\ \vdots \\ Y_{1K} \\ Y_{21} \\ \vdots \\ Y_{2K} \end{bmatrix}, \quad \boldsymbol{\beta} = \begin{bmatrix} \mu_1 \\ \mu_2 \end{bmatrix}, \quad \mathbf{X} = \begin{bmatrix} 1 & 0 \\ 1 & 0 \\ \vdots & \vdots \\ 1 & 0 \\ 0 & 1 \\ \vdots & \vdots \\ 0 & 1 \end{bmatrix} \quad \text{and} \quad \mathbf{e} = \begin{bmatrix} e_{11} \\ e_{12} \\ \vdots \\ e_{1K} \\ e_{21} \\ \vdots \\ e_{2K} \end{bmatrix}.$$

Example 2.2

For the simpler plant growth model

$$Y_{jk} = \mu + e_{jk}, \qquad j = 1, 2 \text{ and } k = 1, \ldots, K$$

so

$$\mathbf{y} = \begin{bmatrix} Y_{11} \\ Y_{12} \\ \vdots \\ Y_{1K} \\ Y_{21} \\ \vdots \\ Y_{2K} \end{bmatrix}, \quad \boldsymbol{\beta} = [\mu], \quad \mathbf{X} = \begin{bmatrix} 1 \\ 1 \\ \\ \vdots \\ \\ 1 \end{bmatrix} \quad \text{and} \quad \mathbf{e} = \begin{bmatrix} e_{11} \\ e_{12} \\ \vdots \\ e_{1K} \\ e_{21} \\ \vdots \\ e_{2K} \end{bmatrix}.$$

Example 2.3

For the model

$$Y_{jk} = \alpha_j + \beta_j x_{jk} + e_{jk}; \quad j = 1, 2 \text{ and } k = 1, \ldots, K$$

for birthweight the corresponding matrix and vector terms are

$$\mathbf{y} = \begin{bmatrix} Y_{11} \\ Y_{12} \\ \vdots \\ Y_{1K} \\ Y_{21} \\ \vdots \\ Y_{2K} \end{bmatrix}, \quad \boldsymbol{\beta} = \begin{bmatrix} \alpha_1 \\ \alpha_2 \\ \beta_1 \\ \beta_2 \end{bmatrix}, \quad \mathbf{X} = \begin{bmatrix} 1 & 0 & x_{11} & 0 \\ 1 & 0 & x_{12} & 0 \\ \vdots & \vdots & \vdots & \vdots \\ 1 & 0 & x_{1K} & 0 \\ 0 & 1 & 0 & x_{21} \\ \vdots & \vdots & \vdots & \vdots \\ 0 & 1 & 0 & x_{2K} \end{bmatrix} \quad \text{and} \quad \mathbf{e} = \begin{bmatrix} e_{11} \\ e_{12} \\ \vdots \\ e_{1K} \\ e_{21} \\ \vdots \\ e_{2K} \end{bmatrix}.$$

Models of the form $\mathbf{y} = \mathbf{X}\boldsymbol{\beta} + \mathbf{e}$ are called *linear models* because the signal part of the model, $\mathbf{X}\boldsymbol{\beta}$, is a linear combination of the parameters and the noise part, $\mathbf{e}$, is also additive. If there are p parameters in the model and N observations, then $\mathbf{y}$ and $\mathbf{e}$ are $N \times 1$ random vectors, $\boldsymbol{\beta}$ is a $p \times 1$ vector of parameters (usually to be estimated) and $\mathbf{X}$ is an $N \times p$ matrix of known constants.

2.5 Exercises

2.1 For the plant growth example verify that the least squares estimators for the parameters in (2.1) and (2.3) are the same as the maximum likelihood estimators.

2.2 Write the equations

$$Y_{jkl} = \mu + \alpha_j + \beta_j x_{jk} + \gamma x_{jk}^2 + e_{jkl},$$

where $j = 1, \ldots, J$, $k = 1, \ldots, K$ and $l = 1, \ldots, L$ in matrix notation. [Hint: form a new independent variable $t_{jk} = x_{jk}^2$.]

2.3 The weights (kg) of 10 people before and after going on a high carbohydrate diet for 3 months are shown in Table 2.6. You want to know if, overall, there was any significant change in weight.

Table 2.6 Weights (kg) of people before and after a diet.

Before	64	71	64	69	76	53	52	72	79	68
After	61	72	63	67	72	49	54	72	74	66

Let Y_{jk} denote the weight of the kth person at time j where $j = 1$ before the diet, $j = 2$ afterwards and $k = 1, \ldots, 10$.

(i) Use the models (2.1) and (2.3) from the plant growth example to test the hypothesis H_0 that there was no change in weight, i.e. $E(Y_{1k}) = E(Y_{2k})$ for all k.

(ii) Let $D_k = Y_{1k} - Y_{2k}$ for $k = 1, \ldots, 10$. If H_0 is true then $E(D_k) = 0$, so another test of H_0 is to compare the models

$$D_k = \mu + e_k \tag{2.7}$$

and

$$D_k = e_k, \tag{2.8}$$

assuming that $e_k \sim \text{NID}(0, \sigma^2)$ for $k = 1, \ldots, 10$ (where μ in (2.7) is not necessarily zero). Use the method of maximum likelihood to estimate μ and compare the values of the likelihood function under (2.7) and (2.8) to test H_0.

(iii) List all the assumptions you made for the analyses in (i) and (ii). Which analysis was more appropriate?

3

EXPONENTIAL FAMILY OF DISTRIBUTIONS AND GENERALIZED LINEAR MODELS

3.1 Introduction

For several decades linear models of the form

$$\mathbf{y} = \mathbf{X}\boldsymbol{\beta}+\mathbf{e} \tag{3.1}$$

with the assumption that the elements of $\mathbf{e}$ are NID$(0, \sigma^2)$ have formed the basis of most analyses of continuous data. For instance the examples in the previous chapter, the comparison of two means (plant growth example) and the relationship between a continuous response variable and a covariate (birthweight example), are both of this form. So, too, are generalizations of these examples to comparisons of more than two means (ANOVA) and the relationship between a continuous response variable and several explanatory variables (multiple regression).

Recent advances in statistical theory and computer software allow us to use methods analogous to those developed for linear models in the following situations:

(i) the response variables have distributions other than the Normal distribution – they may even be categorical rather than continuous;
(ii) the relationship between the response and explanatory variables need not be of the simple linear form in (3.1).

One of these advances has been the recognition that many of the 'nice' properties of the Normal distribution are shared by a wider class of distributions called the *exponential family of distributions*. These are discussed in the next section.

A second advance is the extension of the numerical methods for estimating parameters, from linear combinations like $\mathbf{X}\boldsymbol{\beta}$ in (3.1) to functions of linear combinations such as, $\psi(\mathbf{X}\boldsymbol{\beta})$. In theory the estimation procedures are straightforward. In practice they involve considerable computation so that they have only become feasible with the development of computer programs for numerical optimization of non-linear functions (Chambers, 1973). These are now included in statistical packages such as GLIM (Baker and Nelder, 1978).

This chapter introduces the exponential family of distributions and defines generalized linear models. Methods for parameter estimation and hypothesis testing are developed in Chapters 4 and 5, respectively. Some of the mathematical results are given in the appendices rather than the main text in order to maintain the continuity of the statistical development.

3.2 Exponential family of distributions

Consider a single random variable Y whose probability function, if it is discrete, or probability density function, if it is continuous, depends on a single parameter of interest θ. The distribution belongs to the exponential family if it is of the form

$$f(y;\theta) = s(y)t(\theta)e^{a(y)b(\theta)}, \tag{3.2}$$

where a, b, s and t are known functions. The duality between the random variable and the parameter is emphasized in (3.2) (Barndorff-Nielsen, 1978, Ch. 3). This can be written as

$$f(y;\theta) = \exp\,[a(y)b(\theta)+c(\theta)+d(y)] \tag{3.3}$$

where $s(y) = \exp d(y)$ and $t(\theta) = \exp c(\theta)$.

If $a(y) = y$, (3.3) is said to have the *canonical form* and $b(\theta)$ is sometimes called the *natural parameter* of the distribution.

If there are other parameters in addition to θ they are regarded as *nuisance parameters* forming parts of the functions a, b, c and d, and they are treated as though they are known.

Many well-known distributions belong to the exponential family. For example, the Poisson, Normal and binomial distributions can all be written in the canonical form.

Poisson distribution

$$\begin{aligned} f(y;\lambda) &= \frac{\lambda^y e^{-\lambda}}{y!} \quad \text{for} \quad y = 0, 1, 2, \ldots \\ &= \exp\,[y\log\lambda - \lambda - \log y!]. \end{aligned}$$

Normal distribution, $Y \sim N(\mu, \sigma^2)$

$$\begin{aligned} f(y;\mu) &= \frac{1}{(2\pi\sigma^2)^{\frac{1}{2}}}\exp\left[-\frac{1}{2\sigma^2}(y-\mu)^2\right] \\ &= \exp\left[-\frac{y^2}{2\sigma^2}+\frac{y\mu}{\sigma^2}-\frac{\mu^2}{2\sigma^2}-\tfrac{1}{2}\log(2\pi\sigma^2)\right], \end{aligned}$$

where μ is the parameter of interest and σ^2 is assumed to be known.

Binomial distribution, $Y \sim \mathrm{b}(n, \pi)$

$$f(y;\pi) = \binom{n}{y}\pi^y(1-\pi)^{n-y}, \quad y = 0, 1, \ldots, n$$
$$= \exp\left[y\log\pi - y\log(1-\pi) + n\log(1-\pi) + \log\binom{n}{y}\right].$$

These results are summarized in Table 3.1.

Table 3.1 Poisson, Normal and binomial distributions as members of the exponential family.

Distribution	Natural parameter	c	d
Poisson	$\log\lambda$	$-\lambda$	$-\log y!$
Normal	μ/σ^2	$-\frac{1}{2}\mu^2/\sigma^2 - \frac{1}{2}\log(2\pi\sigma^2)$	$-\frac{1}{2}y^2/\sigma^2$
binomial	$\log\left(\frac{\pi}{1-\pi}\right)$	$n\log(1-\pi)$	$\log\binom{n}{y}$

Other examples of distributions belonging to the exponential family are given in the exercises at the end of the chapter. Not all of them are of the canonical form.

To find the expected value and variance of $a(Y)$ we use the following results which are derived in Appendix 1. If l is a log-likelihood function and $U = \mathrm{d}l/\mathrm{d}\theta$, which is called the *score*, then (A1.2) and (A1.3) give

$$E(U) = 0 \quad \text{and} \quad \mathrm{var}(U) = E(U^2) = E(-U'),$$

where $'$ denotes the derivative with respect to θ.

For the exponential family described by (3.3)

$$l = \log f = a(y)b(\theta) + c(\theta) + d(y),$$

so

$$U = \frac{\mathrm{d}l}{\mathrm{d}\theta} = a(y)b'(\theta) + c'(\theta),$$

and

$$U' = \frac{\mathrm{d}^2 l}{\mathrm{d}\theta^2} = a(y)b''(\theta) + c''(\theta).$$

Thus

$$E(U) = b'(\theta)E[a(Y)] + c'(\theta) = 0,$$

so

$$E[a(Y)] = -c'(\theta)/b'(\theta). \tag{3.4}$$

If $a(Y) = Y$ then $E(Y)$ is denoted by μ, i.e. $\mu = -c'(\theta)/b'(\theta)$.
Also

$$E(-U') = -b''(\theta)E[a(Y)] - c''(\theta).$$

But

$$\mathrm{var}(U) = E(U^2) = [b'(\theta)]^2\,\mathrm{var}[a(Y)] \quad \text{since} \quad E(U) = 0.$$

Now we use the result that $\operatorname{var}(U) = E(-U')$ to obtain

$$\begin{aligned}\operatorname{var}[a(Y)] &= \{-b''(\theta)E[a(Y)] - c''(\theta)\}/[b'(\theta)]^2 \\ &= [b''(\theta)c'(\theta) - c''(\theta)b'(\theta)]/[b'(\theta)]^3. \qquad (3.5)\end{aligned}$$

It is easy to verify (3.4) and (3.5) for the Poisson, Normal and binomial distributions (see Exercise 3.4).

If $Y_1, \ldots, Y_N$ are independent random variables with the same distribution given by (3.3), their joint probability density function is

$$f(y_1, \ldots, y_N; \theta) = \exp\left[b(\theta)\sum_{i=1}^{N} a(y_i) + Nc(\theta) + \sum_{i=1}^{N} d(y_i)\right].$$

Thus $\Sigma a(y_i)$ is a *sufficient* statistic for $b(\theta)$; this means that in a certain sense $\Sigma a(y_i)$ summarizes all the available information about the parameter θ (Cox and Hinkley, 1974, Ch. 2). This result is important for parameter estimation.

Next we consider a class of models based on the exponential family of distributions.

3.3 Generalized linear models

The unity of many statistical methods involving linear combinations of parameters was demonstrated by Nelder and Wedderburn (1972) using the idea of a generalized linear model. This is defined in terms of a set of independent random variables $Y_1, \ldots, Y_N$ each with a distribution from the exponential family with the following properties:

(i) the distribution of each Y_i is of the canonical form and depends on a single parameter θ_i, i.e.

$$f(y_i; \theta_i) = \exp[y_i b_i(\theta_i) + c_i(\theta_i) + d_i(y_i)];$$

(ii) the distributions of all the Y_is are of the same form (e.g. all Normal or all binomial) so that the subscripts on b, c and d are not needed.

Thus the joint probability density function of $Y_i, \ldots, Y_N$ is

$$f(y_i, \ldots, y_N; \theta_1, \ldots, \theta_N) = \exp[\Sigma y_i b(\theta_i) + \Sigma c(\theta_i) + \Sigma d(y_i)]. \qquad (3.6)$$

For model specification, the parameters θ_i are not of direct interest (since there is one for each observation). For a generalized linear model we consider a smaller set of parameters $\beta_1, \ldots, \beta_p$ $(p < N)$ such that a linear combination of the βs is equal to some function of the expected value μ_i of Y_i, i.e.

$$g(\mu_i) = \mathbf{x}_i^{\mathrm{T}} \boldsymbol{\beta}$$

where

(i) g is a monotone, differentiable function called the *link function*;
(ii) $\mathbf{x}_i$ is a $p \times 1$ vector of explanatory variables (covariates and dummy variables for levels of factors);

(iii) $\boldsymbol{\beta}$ is the $p \times 1$ vector of parameters,

$$\boldsymbol{\beta} = [\beta_1, \ldots, \beta_p]^{\mathrm{T}}.$$

Thus a generalized linear model has three components:

(i) response variables $Y_1, \ldots, Y_N$ which are assumed to share the same distribution from the exponential family;
(ii) a set of parameters $\boldsymbol{\beta}$ and explanatory variables

$$\mathbf{X} = [\mathbf{x}_1^{\mathrm{T}}, \ldots, \mathbf{x}_N^{\mathrm{T}}]^{\mathrm{T}};$$

(iii) a link function g such that

$$g(\mu_i) = \mathbf{x}_i^{\mathrm{T}} \boldsymbol{\beta} \quad \text{where } \mu_i = E(Y_i).$$

Such models form the core of this book.

This chapter concludes with two examples of generalized linear models.

Example 3.1

The linear model

$$\mathbf{y} = \mathbf{X}\boldsymbol{\beta} + \mathbf{e}$$

with $\mathbf{e} = [e_1, \ldots, e_N]^{\mathrm{T}}$ and $e_i \sim \mathrm{NID}(0, \sigma^2)$ for $i = 1, \ldots, N$ is a special case of a generalized linear model. This is because the elements Y_i of $\mathbf{y}$ are independent with distributions $\mathrm{N}(\mu_i, \sigma^2)$ where $\mu_i = \mathbf{x}_i^{\mathrm{T}} \boldsymbol{\beta}$ (with $\mathbf{x}_i^{\mathrm{T}}$ denoting the ith row of $\mathbf{X}$). Also the Normal distribution is a member of the exponential family (provided σ^2 is regarded as known). In this case g is the identity function, $g(\mu_i) = \mu_i$.

Example 3.2

A commonly used generalized linear model based on the binomial distribution, $Y_i \sim b(n, \pi_i)$, is obtained by taking the natural parameter as the link function, i.e.

$$\log\left(\frac{\pi_i}{1 - \pi_i}\right) = \mathbf{x}_i^{\mathrm{T}} \boldsymbol{\beta}.$$

Hence

$$\pi_i = \frac{e^{\mathbf{x}_i^{\mathrm{T}} \boldsymbol{\beta}}}{1 + e^{\mathbf{x}_i^{\mathrm{T}} \boldsymbol{\beta}}};$$

this is called the *logistic regression model.* Another possible model for the binomial probabilities is

$$\pi_i = \mathbf{x}_i^{\mathrm{T}} \boldsymbol{\beta}$$

with the identity link function. These are discussed in detail in Chapter 8.

3.4 Exercises

3.1 If the random variable Y has the Gamma distribution with scale parameter θ and known shape parameter ϕ, its probability density function is

$$f(y;\theta) = \frac{y^{\phi-1}\theta^{\phi}\,e^{-y\theta}}{\Gamma(\phi)}.$$

Show that this distribution belongs to the exponential family and hence find the natural parameter, $E(Y)$ and $\text{var}(Y)$.

3.2 Show that the following probability distributions belong to the exponential family:

(i) $f(y;\theta) = \theta y^{-\theta-1}$ (Pareto distribution)
(ii) $f(y;\theta) = \theta e^{-y\theta}$ (exponential distribution)
(iii) $f(y;\theta) = \binom{y+r-1}{r-1}\theta^r(1-\theta)^y$, where r is known (negative binomial distribution).

3.3 For the binomial distribution show from first principles that

$$E(U) = 0 \quad \text{and} \quad \text{var}(U) = E(U^2) = E(-U')$$

where $U = dl/d\theta$ and l is the log-likelihood function.

3.4 Use (3.4) and (3.5) to verify these results:

(i) for the Poisson distribution, $E(Y) = \text{var}(Y)$;
(ii) if $Y \sim N(\mu, \sigma^2)$, $E(Y) = \mu$ and $\text{var}(Y) = \sigma^2$;
(iii) if $Y \sim b(n, \pi)$, $E(Y) = n\pi$ and $\text{var}(Y) = n\pi(1-\pi)$.

4 ESTIMATION

4.1 Introduction

Two of the most commonly used approaches to the statistical estimation of parameters are the method of maximum likelihood and the method of least squares. This chapter begins by reviewing the principle of each of these methods and some properties of the estimators. Then the method of maximum likelihood is used for generalized linear models. Usually the estimates have to be obtained numerically by an iterative procedure which turns out to be closely related to weighted least squares estimation.

In the next chapter we consider the distributional properties of estimators for generalized linear models, including the calculation of standard errors and confidence regions.

4.2 Method of maximum likelihood

Let $Y_1, \ldots, Y_N$ be N random variables with the joint probability density function $f(y_1, \ldots, y_N; \theta_1, \ldots, \theta_p)$ which depends on parameters $\theta_1, \ldots, \theta_p$. For brevity we denote $[y_1, \ldots, y_N]^{\mathrm{T}}$ by $\mathbf{y}$ and $[\theta_1, \ldots, \theta_p]^{\mathrm{T}}$ by $\boldsymbol{\theta}$.

Algebraically the *likelihood function* $L(\boldsymbol{\theta}; \mathbf{y})$ is the same as $f(\mathbf{y}; \boldsymbol{\theta})$ but the change in notation reflects a shift of emphasis from the random variables $\mathbf{Y}$, with $\boldsymbol{\theta}$ fixed, to the parameters $\boldsymbol{\theta}$ with $\mathbf{y}$ fixed (specifically $\mathbf{y}$ represents the observations). Let Ω denote the parameter space, i.e. all possible values of the parameter vector $\boldsymbol{\theta}$. The *maximum likelihood estimator* of $\boldsymbol{\theta}$ is defined as the vector $\hat{\boldsymbol{\theta}}$ such that

$$L(\hat{\boldsymbol{\theta}}; \mathbf{y}) \geqslant L(\boldsymbol{\theta}; \mathbf{y}) \qquad \text{for all } \boldsymbol{\theta} \in \Omega.$$

Equivalently, if $l(\boldsymbol{\theta}; \mathbf{y}) = \log L(\boldsymbol{\theta}; \mathbf{y})$ is the *log-likelihood function*, then $\hat{\boldsymbol{\theta}}$ is the maximum likelihood estimator if

$$l(\hat{\boldsymbol{\theta}}; \mathbf{y}) \geqslant l(\boldsymbol{\theta}; \mathbf{y}) \qquad \text{for all } \boldsymbol{\theta} \in \Omega.$$

The most convenient way to obtain the maximum likelihood estimator is to examine all the local maxima of $l(\boldsymbol{\theta}; \mathbf{y})$. These are

(i) the solutions of

$$\frac{\partial l(\boldsymbol{\theta};\mathbf{y})}{\partial \theta_j} = 0, \qquad j = 1, \ldots, p$$

such that $\boldsymbol{\theta}$ belongs to Ω and the matrix of second derivatives

$$\frac{\partial^2 l(\boldsymbol{\theta};\mathbf{y})}{\partial \theta_j \partial \theta_k}$$

is negative definite; and

(ii) any values of $\boldsymbol{\theta}$ at the edges of the parameter space Ω which correspond to maxima of $l(\boldsymbol{\theta};\mathbf{y})$.

The value $\hat{\boldsymbol{\theta}}$ giving the largest of the local maxima is the maximum likelihood estimator. For models considered in this book there is usually a unique maximum given by $\partial L/\partial \boldsymbol{\theta} = \mathbf{0}$.

An important property of maximum likelihood estimators is that if $\psi(\boldsymbol{\theta})$ is any function of the parameters $\boldsymbol{\theta}$, then the maximum likelihood estimator of ψ is

$$\hat{\psi} = \psi(\hat{\boldsymbol{\theta}}).$$

This follows from the definition of $\hat{\boldsymbol{\theta}}$. It is sometimes called the *invariance property* of maximum likelihood estimators. A consequence is that we can work with the parameterization of a model which is most convenient for maximum likelihood estimation and then use the invariance property to obtain maximum likelihood estimates for other parameterizations.

Other properties of maximum likelihood estimators include consistency, sufficiency and asymptotic efficiency. These are discussed in detail in books on theoretical statistics, for example Cox and Hinkley (1974), Ch. 9.

4.3 Method of least squares

Let $Y_1, \ldots, Y_N$ be random variables with expected values

$$E(Y_i) = \mu_i = \mu_i(\boldsymbol{\beta}), \qquad i = 1, \ldots, N,$$

where $\boldsymbol{\beta} = [\beta_1, \ldots, \beta_p]^{\mathrm{T}}$ $(p < N)$ are the parameters to be estimated. Consider the formulation

$$Y_i = \mu_i + e_i, \qquad i = 1, \ldots, N,$$

in which μ_i represents the 'signal' component of Y_i and e_i represents the 'noise' component. The *method of least squares* consists of finding estimators $\hat{\boldsymbol{\beta}}$ which minimize the sum of squares of the error terms

$$S = \sum_{i=1}^{N} e_i^2 = \sum_{i=1}^{N} [Y_i - \mu_i(\boldsymbol{\beta})]^2. \tag{4.1}$$

In matrix notation this is

$$S = (\mathbf{y}-\boldsymbol{\mu})^{\mathrm{T}}(\mathbf{y}-\boldsymbol{\mu}),$$

where $\mathbf{y} = [Y_1, \ldots, Y_N]^{\mathrm{T}}$ and $\boldsymbol{\mu} = [\mu_1, \ldots, \mu_N]^{\mathrm{T}}$. Generally the estimator $\hat{\boldsymbol{\beta}}$ is obtained by differentiating S with respect to each element β_j of $\boldsymbol{\beta}$ and solving the simultaneous equations

$$\frac{\partial S}{\partial \beta_j} = 0, \qquad j = 1, \ldots, p.$$

Of course it is necessary to check that the solutions correspond to minima (i.e. the matrix of second derivatives is positive definite) and to identify the global minimum from among these solutions and any local minima at the boundary of the parameter space.

In practice there may be additional information about the Y_is, for example that some observations are less reliable (i.e. have larger variance) than others. In such a case we may wish to weight the terms in (4.1) accordingly and minimize the sum

$$S_w = \sum_{i=1}^{N} w_i(Y_i-\mu_i)^2.$$

where the terms w_i represent weights, e.g. $w_i = [\mathrm{var}(Y_i)]^{-1}$. More generally the Y_is may be correlated; let $\mathbf{V}$ denote their variance–covariance matrix. Then *weighted least squares* estimators are obtained by minimizing

$$S_w = (\mathbf{y}-\boldsymbol{\mu})^{\mathrm{T}}\mathbf{V}^{-1}(\mathbf{y}-\boldsymbol{\mu}).$$

In particular if the terms μ_i are linear combinations of parameters $\beta_j (j = 1, \ldots, p < N)$, i.e. $\boldsymbol{\mu} = \mathbf{X}\boldsymbol{\beta}$ for some $N \times p$ matrix $\mathbf{X}$, then

$$S_w = (\mathbf{y}-\mathbf{X}\boldsymbol{\beta})^{\mathrm{T}}\mathbf{V}^{-1}(\mathbf{y}-\mathbf{X}\boldsymbol{\beta}). \tag{4.2}$$

For (4.2) the derivatives of S_w with respect to the elements β_j of $\boldsymbol{\beta}$ are

$$\frac{\partial S_w}{\partial \boldsymbol{\beta}} = -2\mathbf{X}^{\mathrm{T}}\mathbf{V}^{-1}(\mathbf{y}-\mathbf{X}\boldsymbol{\beta}),$$

so the weighted least squares estimator of $\boldsymbol{\beta}$ is the solution of the *normal equations*

$$\mathbf{X}^{\mathrm{T}}\mathbf{V}^{-1}\mathbf{X}\mathbf{b} = \mathbf{X}^{\mathrm{T}}\mathbf{V}^{-1}\mathbf{y} \tag{4.3}$$

(since it can be shown that the matrix of second derivatives is positive definite).

An important distinction between the methods of least squares and maximum likelihood is that the former can be used without making assumptions about the distributions of the response variables Y_i beyond specifying their expectations and possibly their variance–covariance structure. In contrast, to obtain maximum likelihood estimators we need to specify the joint probability distribution of the Y_is. However, to obtain the sampling distribution of the least squares estimators $\mathbf{b}$ additional assumptions about the Y_is are generally required. Thus in practice there is little advantage in using the method of least squares unless the estimation equations are computationally simpler.

4.4 Estimation for generalized linear models

We wish to obtain maximum likelihood estimators of the parameters $\boldsymbol{\beta}$ for the generalized linear models defined in Section 3.3: the log-likelihood function for independent responses $Y_1, \ldots, Y_N$ is

$$l(\boldsymbol{\theta}; \mathbf{y}) = \Sigma y_i b(\theta_i) + \Sigma c(\theta_i) + \Sigma d(y_i)$$

where

$$E(Y_i) = \mu_i = -c'(\theta_i)/b'(\theta_i),$$

and

$$g(\mu_i) = \mathbf{x}_i^{\mathrm{T}} \boldsymbol{\beta} = \eta_i$$

where g is monotone and differentiable.

A property of the exponential family of distributions is that they satisfy enough regularity conditions to ensure that the global maximum of l is given uniquely by the solution of $\partial l/\partial \boldsymbol{\theta} = \mathbf{0}$, or equivalently, $\partial l/\partial \boldsymbol{\beta} = \mathbf{0}$ (see Cox and Hinkley, 1974, Ch. 9).

In Appendix 2 (A2.6), it is shown that

$$\frac{\partial l}{\partial \beta_j} = U_j = \sum_{i=1}^{N} \frac{(y_i - \mu_i) x_{ij}}{\operatorname{var}(Y_i)} \left(\frac{\partial \mu_i}{\partial \eta_i}\right) \tag{4.4}$$

where x_{ij} is the jth element of $\mathbf{x}_i^{\mathrm{T}}$. In general the equations $U_j = 0, j = 1, \ldots, p$ are non-linear and they have to be solved numerically by iteration. For the *Newton–Raphson method* the mth approximation is given by

$$\mathbf{b}^{(m)} = \mathbf{b}^{(m-1)} - \left[\frac{\partial^2 l}{\partial \beta_j \partial \beta_k}\right]^{-1}_{\boldsymbol{\beta} = \mathbf{b}^{(m-1)}} \mathbf{U}^{(m-1)} \tag{4.5}$$

where

$$\left[\frac{\partial^2 l}{\partial \beta_j \partial \beta_k}\right]_{\boldsymbol{\beta} = \mathbf{b}^{(m-1)}}$$

is the matrix of second derivatives of l evaluated at $\boldsymbol{\beta} = \mathbf{b}^{(m-1)}$ and $\mathbf{U}^{(m-1)}$ is the vector of first derivatives $U_j = \partial l/\partial \beta_j$ evaluated at $\boldsymbol{\beta} = \mathbf{b}^{(m-1)}$. (This is the multidimensional analogue of the Newton–Raphson method for finding a solution of the equation $f(x) = 0$, namely

$$x^{(m)} = x^{(m-1)} - f(x^{(m-1)})/f'(x^{(m-1)}).)$$

An alternative procedure which is sometimes simpler than the Newton–

Raphson method is called the *method of scoring*. It involves replacing the matrix of second derivatives in (4.5) by the matrix of expected values

$$E\left[\frac{\partial^2 l}{\partial\beta_j\,\partial\beta_k}\right].$$

In Appendix 1 it is shown that this is equal to the negative of the *information matrix*, the variance–covariance matrix of the U_js, $\mathscr{I} = E[\mathbf{U}\mathbf{U}^{\mathrm{T}}]$ which has elements

$$\mathscr{I}_{jk} = E[U_j U_k] = E\left[\frac{\partial l}{\partial\beta_j}\frac{\partial l}{\partial\beta_k}\right] = -E\left[\frac{\partial^2 l}{\partial\beta_j\,\partial\beta_k}\right].$$

Thus (4.5) is replaced by

$$\mathbf{b}^{(m)} = \mathbf{b}^{(m-1)} + [\mathscr{I}^{(m-1)}]^{-1}\mathbf{U}^{(m-1)} \tag{4.6}$$

where $\mathscr{I}^{(m-1)}$ denotes the information matrix evaluated at $\mathbf{b}^{(m-1)}$. Multiplication by $\mathscr{I}^{(m-1)}$ in (4.6) gives

$$\mathscr{I}^{(m-1)}\mathbf{b}^{(m)} = \mathscr{I}^{(m-1)}\mathbf{b}^{(m-1)} + \mathbf{U}^{(m-1)}. \tag{4.7}$$

For generalized linear models the (j, k)th element of $\mathscr{I}$ is

$$\mathscr{I}_{jk} = \sum_{i=1}^{N} \frac{x_{ij}x_{ik}}{\operatorname{var}(Y_i)}\left(\frac{\partial\mu_i}{\partial\eta_i}\right)^2 \tag{4.8}$$

(see Appendix 2, (A2.7)). Thus $\mathscr{I}$ can be written as

$$\mathscr{I} = \mathbf{X}^{\mathrm{T}}\mathbf{W}\mathbf{X}$$

where $\mathbf{W}$ is the $N \times N$ diagonal matrix with elements

$$w_{ii} = \frac{1}{\operatorname{var}(Y_i)}\left(\frac{\partial\mu_i}{\partial\eta_i}\right)^2. \tag{4.9}$$

The expression on the right-hand side of (4.7) is the vector with elements

$$\sum_k\sum_i \frac{x_{ij}x_{ik}}{\operatorname{var}(Y_i)}\left(\frac{\partial\mu_i}{\partial\eta_i}\right)^2 b_k^{(m-1)} + \sum_i \frac{(y_i-\mu_i)x_{ij}}{\operatorname{var}(Y_i)}\left(\frac{\partial\mu_i}{\partial\eta_i}\right),$$

evaluated at $\mathbf{b}^{(m-1)}$; this follows from (4.8) and (4.4). Thus the right-hand side of (4.7) can be written as

$$\mathscr{I}^{(m-1)}\mathbf{b}^{(m-1)} + \mathbf{U}^{(m-1)} = \mathbf{X}^{\mathrm{T}}\mathbf{W}\mathbf{z}$$

where $\mathbf{z}$ has elements

$$z_i = \sum_k x_{ik}b_k^{(m-1)} + (y_i-\mu_i)\left(\frac{\partial\eta_i}{\partial\mu_i}\right) \tag{4.10}$$

with μ_i and $\partial\eta_i/\partial\mu_i$ evaluated at $\mathbf{b}^{(m-1)}$.

Hence the iterative equation for the method of scoring, (4.7), can be written as

$$\mathbf{X}^{\mathrm{T}}\mathbf{W}\mathbf{X}\mathbf{b}^{(m)} = \mathbf{X}^{\mathrm{T}}\mathbf{W}\mathbf{z}. \tag{4.11}$$

This has the same form as the normal equations for a linear model obtained by weighted least squares, (4.3), except that (4.11) has to be solved iteratively because in general $\mathbf{z}$ and $\mathbf{W}$ depend on $\mathbf{b}$. Thus for generalized linear models maximum likelihood estimators are obtained by an *iterative weighted least squares* procedure.

Usually a computer is needed to solve (4.11). Most statistical packages which include analyses based on generalized linear models have efficient programs for calculating the solutions. They begin by using some initial approximation $\mathbf{b}^{(0)}$ to evaluate $\mathbf{z}$ and $\mathbf{W}$, then (4.11) is solved to give $\mathbf{b}^{(1)}$ which in turn is used to obtain better approximations for $\mathbf{z}$ and $\mathbf{W}$, and so on until adequate convergence is achieved. When the difference between successive approximations $\mathbf{b}^{(m)}$ and $\mathbf{b}^{(m-1)}$ is sufficiently small, $\mathbf{b}^{(m)}$ is taken as the maximum likelihood estimate. The example below illustrates the use of this estimation procedure.

A slightly different derivation of (4.11) can be obtained using a Taylor series approximation of the log-likelihood function (see Exercise 4.4).

Example 4.1 *Simple linear regression for Poisson responses*

The data in Table 4.1 are counts y_i observed at various values of a covariate x. They are plotted in Fig. 4.1.

Table 4.1 Poisson regression data.

y_i	2	3	6	7	8	9	10	12	15
x_i	−1	−1	0	0	0	0	1	1	1

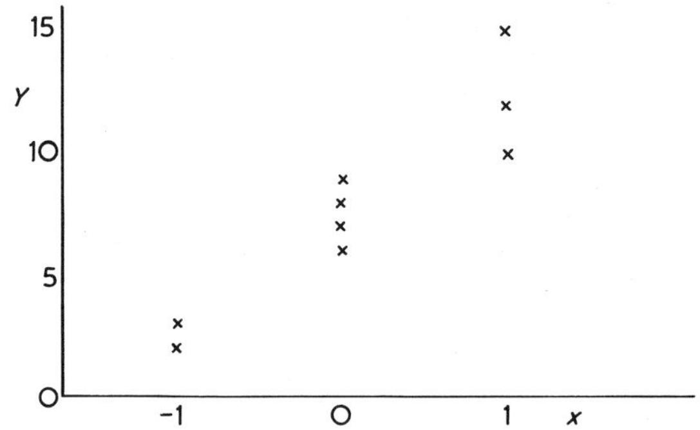

Figure 4.1 Plot of data in Table 4.1.

Either on substantive grounds or from observing that the variability increases with x, let us assume that the responses Y_i are Poisson random variables with

$$E(Y_i) = \text{var}(Y_i) = \mu_i = \beta_1 + \beta_2 x_i.$$

To fit this model by the methods described above we use $\boldsymbol{\beta} = [\beta_1 \beta_2]^T$ and $\mathbf{x}_i^T = [1\ x_i]$ for $i = 1, \ldots, N = 9$. Hence $z_i = y_i$ (because the link function is the identity so $\partial\mu_i/\partial\eta_i = 1$),

$$\mathscr{I} = \mathbf{X}^T\mathbf{W}\mathbf{X} = \begin{bmatrix} \Sigma(\beta_1+\beta_2 x_i)^{-1} & \Sigma x_i(\beta_1+\beta_2 x_i)^{-1} \\ \Sigma x_i(\beta_1+\beta_2 x_i)^{-1} & \Sigma x_i^2(\beta_1+\beta_2 x_i)^{-1} \end{bmatrix}$$

and

$$\mathbf{X}^T\mathbf{W}\mathbf{z} = \begin{bmatrix} \Sigma\, y_i(\beta_1+\beta_2 x_i)^{-1} \\ \Sigma\, x_i y_i(\beta_1+\beta_2 x_i)^{-1} \end{bmatrix}.$$

The maximum likelihood estimates are obtained iteratively from the equations

$$(\mathbf{X}^T\mathbf{W}\mathbf{X})^{(m-1)}\mathbf{b}^{(m)} = (\mathbf{X}^T\mathbf{W}\mathbf{z})^{(m-1)}$$

where the superscript $(m-1)$ denotes evaluation at $\mathbf{b}^{(m-1)}$. From Fig. 4.1 we choose initial values $b_1^{(0)} = 7$ and $b_2^{(0)} = 5$. Successive approximations are shown in the Table 4.2. Thus the maximum likelihood estimates, correct to 4 decimal places, are $b_1 = 7.4516$ and $b_2 = 4.9353$.

Table 4.2 Successive approximations for regression coefficients

m	0	1	2	3
$b_1^{(m)}$	7	7.450	7.4516	7.4516
$b_2^{(m)}$	5	4.937	4.9353	4.9353

4.5 Exercises

4.1 For the data in Example 4.1 fit the model with the Y_is as Poisson variables so that $E(Y_i) = \text{var}(Y_i) = \mu_i$ and

$$\log\mu_i = \beta_1 + \beta_2 x_i$$

(i.e. this choice of link function is based on the natural parameter of the distribution).

4.2 Consider a random sample $Y_1, \ldots, Y_N$ with $Y_i \sim \text{N}(\log\beta, \sigma^2)$ where σ^2 is known. Verify the results in Section 4.4 by finding the maximum likelihood estimator of β
 (i) by first principles, and
 (ii) using (4.11).

4.3 Let $Y_1, \ldots, Y_N$ be independent random variables with $Y_i \sim \text{N}(\mathbf{x}_i^T\boldsymbol{\beta}, \sigma_i^2)$. Show that the maximum likelihood estimator of $\boldsymbol{\beta}$ is the solution of $\mathbf{X}^T\mathbf{V}^{-1}\mathbf{X}\mathbf{b} = \mathbf{X}^T\mathbf{V}^{-1}\mathbf{y}$ where $\mathbf{V}$ is the diagonal matrix with elements $v_{ii} = \sigma_i^2$. (Since this is the same as (4.3), for linear models with Normal errors maximum likelihood estimators and least squares estimators are identical.)

4.4 The second-order Taylor series approximation of the log-likelihood function $l(\boldsymbol{\beta}; \mathbf{y})$ obtained by expanding about $\boldsymbol{\beta} = \boldsymbol{\beta}^*$ is

$$l(\boldsymbol{\beta}; \mathbf{y}) = l(\boldsymbol{\beta}^*; \mathbf{y}) + (\boldsymbol{\beta} - \boldsymbol{\beta}^*)^T \mathbf{U} + \tfrac{1}{2}(\boldsymbol{\beta} - \boldsymbol{\beta}^*)^T \mathbf{H}(\boldsymbol{\beta} - \boldsymbol{\beta}^*)$$

where $\mathbf{U}$, the $p \times 1$ vector with elements $U_j = \partial l / \partial \beta_j$ and $\mathbf{H}$, the $p \times p$ matrix

$$\left[\frac{\partial^2 l}{\partial \beta_j \partial \beta_k}\right],$$

are evaluated at $\boldsymbol{\beta}^*$. Show that the maximum likelihood estimator $\hat{\boldsymbol{\beta}}$ obtained from this approximation is the same as the solution of the Newton–Raphson equation (4.5).

5
INFERENCE

5.1 Introduction

As outlined in Chapter 1 and illustrated in Chapter 2, statistical modelling involves three steps:

(i) specifying models;
(ii) estimating parameters;
(iii) making inferences, that is finding confidence intervals, assessing the goodness of fit of models and testing hypotheses.

For generalized linear models, model specification and parameter estimation are discussed in Chapters 3 and 4. This chapter covers the third step. It describes the derivation and use of sampling distributions for parameter estimators and for the statistics used for measuring goodness of fit.

In the particular case of linear models with Normally distributed error terms, the sampling distributions can be determined exactly. In general the problem of finding exact distributions is intractable and we rely instead on large-sample, asymptotic results. The rigorous development of these results requires careful attention to various regularity conditions. For independent observations from the exponential family of distributions, and in particular for generalized linear models, the necessary conditions are satisfied. We consider only the major steps and not the finer points involved in deriving the sampling distributions. For a more detailed discussion see, for example, Cox and Hinkley (1974) Ch. 9.

We assume, for simplicity, that for any parameter $\boldsymbol{\theta}$ of length p, the estimator $\hat{\boldsymbol{\theta}}$ has variance–covariance matrix $\mathbf{V}$ which is non-singular. Therefore, at least asymptotically,

$$(\hat{\boldsymbol{\theta}}-\boldsymbol{\theta})^{\mathrm{T}}\mathbf{V}^{-1}(\boldsymbol{\theta}-\boldsymbol{\theta}) \sim \chi^2_p$$

provided that $E(\hat{\boldsymbol{\theta}}) = \boldsymbol{\theta}$. If $\mathbf{V}$ is singular with rank $q < p$, then $(\hat{\boldsymbol{\theta}}-\boldsymbol{\theta})^{\mathrm{T}}\mathbf{V}^{-}(\hat{\boldsymbol{\theta}}-\boldsymbol{\theta}) \sim \chi^2_q$ where $\mathbf{V}^{-}$ is a generalized inverse; alternatively the model might be re-expressed in terms of a parameter vector $\boldsymbol{\psi}$ of length q with non-singular variance–covariance matrix $\mathbf{W}$ so that $(\hat{\boldsymbol{\psi}}-\boldsymbol{\psi})^{\mathrm{T}}\mathbf{W}^{-1}(\hat{\boldsymbol{\psi}}-\boldsymbol{\psi}) \sim \chi^2_q$.

We begin with the sampling distribution for the scores $U_j = \partial l/\partial\beta_j$. Then we derive the distribution of the maximum likelihood estimator **b** and hence obtain confidence regions for the parameters $\boldsymbol{\beta}$. A goodness of fit statistic is derived from the likelihood ratio test and its sampling distribution is obtained. Finally, the use of this statistic for hypothesis testing is discussed.

5.2 Sampling distribution for scores

For generalized linear models the score with respect to parameter β_j is

$$U_j = \frac{\partial l}{\partial \beta_j}, \quad j = 1, 2, \ldots, p,$$

where l is the log-likelihood function and the elements of $\boldsymbol{\beta} = [\beta_1, \ldots, \beta_p]^{\mathrm{T}}$ are the parameters in the linear component of the model.

For $\mathbf{U} = [U_1, \ldots, U_p]^{\mathrm{T}}$, in Appendix 1 it is shown that

$$E(\mathbf{U}) = \mathbf{0} \qquad \text{and} \qquad E(\mathbf{U}\mathbf{U}^{\mathrm{T}}) = \mathscr{I}$$

where $\mathscr{I}$ is the information matrix. By the Central Limit Theorem the asymptotic distribution of **U** is the multivariate Normal distribution $\mathrm{N}(\mathbf{0}, \mathscr{I})$. Hence, by the definition of the central chi-squared distribution (1.2), for large samples

$$\mathbf{U}^{\mathrm{T}}\mathscr{I}^{-1}\mathbf{U} \sim \chi^2_p. \tag{5.1}$$

(provided that $\mathscr{I}$ is non-singular so $\mathscr{I}^{-1}$ exists).

5.3 Sampling distribution for maximum likelihood estimators

Suppose that the log-likelihood function has a unique maximum at **b** and that this estimator **b** is near the true value of the parameter $\boldsymbol{\beta}$. The first order Taylor approximation for the score vector $\mathbf{U}(\boldsymbol{\beta})$ about **b** is

$$\mathbf{U}(\boldsymbol{\beta}) \simeq \mathbf{U}(\mathbf{b}) + \mathbf{H}(\mathbf{b})(\boldsymbol{\beta} - \mathbf{b})$$

where $\mathbf{H}(\mathbf{b})$ denotes the matrix of second derivatives

$$\left[\frac{\partial^2 l}{\partial \beta_j \partial \beta_k}\right]$$

evaluated at **b**. Asymptotically **H** is equal to its expected value which, from Appendix 1, is related to the information matrix by

$$\mathscr{I} = E(\mathbf{U}\mathbf{U}^{\mathrm{T}}) = E(-\mathbf{H}).$$

Therefore, for large samples

$$\mathbf{U}(\boldsymbol{\beta}) \simeq \mathbf{U}(\mathbf{b}) - \mathscr{I}(\boldsymbol{\beta} - \mathbf{b}).$$

But $\mathbf{U}(\mathbf{b}) = \mathbf{0}$ by the definition of $\mathbf{b}$ so approximately

$$(\mathbf{b}-\boldsymbol{\beta}) \simeq \mathscr{I}^{-1}\mathbf{U},$$

provided $\mathscr{I}$ is non-singular. If $\mathscr{I}$ is regarded as fixed then

$$E(\mathbf{b}-\boldsymbol{\beta}) \simeq \mathscr{I}^{-1}E(\mathbf{U}) = \mathbf{0}$$

because $E(\mathbf{U}) = \mathbf{0}$. Similarly

$$E[(\mathbf{b}-\boldsymbol{\beta})(\mathbf{b}-\boldsymbol{\beta})^{\mathrm{T}}] \simeq \mathscr{I}^{-1}E(\mathbf{U}\mathbf{U}^{\mathrm{T}})\mathscr{I}^{-1} = \mathscr{I}^{-1}$$

because $\mathscr{I} = E(\mathbf{U}\mathbf{U}^{\mathrm{T}})$ and it is symmetric.

Thus for large samples

$$\mathbf{b}-\boldsymbol{\beta} \sim \mathrm{N}(\mathbf{0}, \mathscr{I}^{-1}) \tag{5.2}$$

and hence

$$(\mathbf{b}-\boldsymbol{\beta})^{\mathrm{T}}\mathscr{I}(\mathbf{b}-\boldsymbol{\beta}) \sim \chi^2_p. \tag{5.3}$$

If $\mathscr{I}$ depends on $\boldsymbol{\beta}$, for practical applications we often use $\mathscr{I}(\mathbf{b})$ or even $-\mathbf{H}(\mathbf{b})$ in (5.2) or (5.3).

For linear models with Normally distributed error terms results (5.2) and (5.3) are exact rather than asymptotic. This is shown in the following example.

Example 5.1

Suppose that the response variables $Y_1, \ldots, Y_N$ are independently distributed with $Y_i \sim \mathrm{N}(\mathbf{x}_i^{\mathrm{T}}\boldsymbol{\beta}, \sigma^2)$ and that $\mathbf{X}^{\mathrm{T}}\mathbf{X}$ is non-singular, where $\mathbf{X}$ is the $N \times p$ matrix with rows $\mathbf{x}_i^{\mathrm{T}}$. In this case $E(Y_i) = \mu_i = \mathbf{x}_i^{\mathrm{T}}\boldsymbol{\beta} = \eta_i$ so $\partial\mu_i/\partial\eta_i = 1$ and hence from the previous chapter:

(i) the elements of $\mathscr{I}$ are

$$\mathscr{I}_{jk} = \frac{1}{\sigma^2}\sum_{i=1}^{N} x_{ij}x_{ik} \qquad \text{(from (4.8))}$$

so that

$$\mathscr{I} = \frac{1}{\sigma^2}\mathbf{X}^{\mathrm{T}}\mathbf{X}; \tag{5.4}$$

(ii) $\mathbf{W}$ is the diagonal matrix with all elements equal to $1/\sigma^2$ (from (4.9));

(iii) $\mathbf{z} = \mathbf{X}\mathbf{b}+\mathbf{y}-\mathbf{X}\mathbf{b} = \mathbf{y}$ (from (4.10));

(iv) the maximum likelihood estimator $\mathbf{b}$ is the solution of $\mathbf{X}^{\mathrm{T}}\mathbf{X}\mathbf{b} = \mathbf{X}^{\mathrm{T}}\mathbf{y}$ (from (4.11)); i.e.

$$\mathbf{b} = (\mathbf{X}^{\mathrm{T}}\mathbf{X})^{-1}\mathbf{X}^{\mathrm{T}}\mathbf{y}.$$

Thus $\mathbf{b}$ is a linear combination of Normally distributed random variables $Y_1, \ldots, Y_N$ so it, too, is Normally distributed.

Also $\mathbf{b}$ is an unbiased estimator because

$$E(\mathbf{b}) = (\mathbf{X}^{\mathrm{T}}\mathbf{X})^{-1}\mathbf{X}^{\mathrm{T}}E(\mathbf{y}) = (\mathbf{X}^{\mathrm{T}}\mathbf{X})^{-1}\mathbf{X}^{\mathrm{T}}\mathbf{X}\boldsymbol{\beta} = \boldsymbol{\beta}.$$

To obtain the variance–covariance matrix for $\mathbf{b}$ we use

$$\mathbf{b}-\boldsymbol{\beta} = (\mathbf{X}^{\mathrm{T}}\mathbf{X})^{-1}\mathbf{X}^{\mathrm{T}}\mathbf{y}-\boldsymbol{\beta} = (\mathbf{X}^{\mathrm{T}}\mathbf{X})^{-1}\mathbf{X}^{\mathrm{T}}(\mathbf{y}-\mathbf{X}\boldsymbol{\beta}),$$

therefore

$$E[(\mathbf{b}-\boldsymbol{\beta})(\mathbf{b}-\boldsymbol{\beta})^{\mathrm{T}}] = (\mathbf{X}^{\mathrm{T}}\mathbf{X})^{-1}\mathbf{X}^{\mathrm{T}}E[(\mathbf{y}-\mathbf{X}\boldsymbol{\beta})(\mathbf{y}-\mathbf{X}\boldsymbol{\beta})^{\mathrm{T}}]\mathbf{X}(\mathbf{X}^{\mathrm{T}}\mathbf{X})^{-1}$$
$$= \sigma^2(\mathbf{X}^{\mathrm{T}}\mathbf{X})^{-1} = \mathscr{I}^{-1}$$

since $E[(\mathbf{y}-\mathbf{X}\boldsymbol{\beta})(\mathbf{y}-\mathbf{X}\boldsymbol{\beta})^{\mathrm{T}}]$ is the diagonal matrix with elements σ^2 and by (5.4)

$$\mathscr{I} = \frac{1}{\sigma^2}\mathbf{X}^{\mathrm{T}}\mathbf{X}.$$

Thus the exact distribution of $\mathbf{b}$ is $\mathrm{N}(\boldsymbol{\beta}, \mathscr{I}^{-1})$.

5.4 Confidence intervals for the model parameters

For the above example, the exact distribution of $\mathbf{b}$, $\mathrm{N}(\boldsymbol{\beta}, \mathscr{I}^{-1})$, can be used to calculate confidence regions for $\boldsymbol{\beta}$. For instance if σ^2 is known, a 95% confidence interval for β_j is

$$b_j \pm 1.96\,(v_{jj})^{\frac{1}{2}}$$

where v_{jj} is the (j,j)th element of $\mathscr{I}^{-1} = \sigma^2(\mathbf{X}^{\mathrm{T}}\mathbf{X})^{-1}$.

In general, $\mathscr{I}$ may depend on $\boldsymbol{\beta}$ and it can be estimated by substituting $\mathbf{b}$ for $\boldsymbol{\beta}$. Let v_{jk} denote the (j, k)th element of $[\mathscr{I}(\mathbf{b})]^{-1}$. Since the distributional results usually depend on having a large sample, the standard errors $(v_{jj})^{\frac{1}{2}}$ and correlation coefficients

$$r_{jk} = \frac{v_{jk}}{(v_{jj})^{\frac{1}{2}}\,(v_{kk})^{\frac{1}{2}}}$$

provide informal rather than exact guides to the reliability and interdependence of the b_js. Also they can be used to calculate approximate confidence intervals.

Example 5.2

In Example 4.1 we fitted a model involving Poisson distributed responses Y_i with $E(Y_i) = \beta_1 + \beta_2 x_i$ to the data shown in Table 5.1.

Table 5.1 Poisson regression data.

y_i	2	3	6	7	8	9	10	12	15
x_i	−1	−1	0	0	0	0	1	1	1

The maximum likelihood estimates are $b_1 = 7.4516$ and $b_2 = 4.9353$. The inverse of the information matrix evaluated at $\mathbf{b}$ is

$$\mathscr{I}^{-1} = \begin{bmatrix} 0.7817 & 0.4166 \\ 0.4166 & 1.1863 \end{bmatrix}.$$

This shows that b_2 is somewhat less reliable than b_1. The correlation coefficient for b_1 and b_2 is

$$r = \frac{0.4166}{(0.7817)^{\frac{1}{2}}(1.1863)^{\frac{1}{2}}} \simeq 0.43.$$

Also, for example, an approximate 95% confidence interval for β_1 is given by $7.4516 \pm 1.96(0.7817)^{\frac{1}{2}}$, i.e. (5.72, 9.18).

5.5 Comparison of models

Hypotheses about $\boldsymbol{\beta}$ can be tested using the sampling distribution of $\mathbf{b}$ (result (5.2)). An alternative approach consists of specifying each hypothesis in terms of a model and comparing measures of goodness of fit for each model.

We only consider generalized linear models which

(i) are based on the same distribution from the exponential family,
(ii) have the same link function, but
(iii) differ in the numbers of parameters used.

The adequacy of a model is defined relative to a *maximal* (or *saturated*) *model* which has the same number of parameters as observations and so provides a complete description of the data (at least for this distribution). The maximal model involves parameters

$$\boldsymbol{\beta}_{\max} = [\beta_1, \ldots, \beta_N]^T,$$

where N is the number of observations. We compare this with another model specified by a parameter vector $\boldsymbol{\beta}$ of length p with $p < N$, i.e. $\boldsymbol{\beta} = [\beta_1, \ldots, \beta_p]^T$.

In terms of the likelihood function $L(\boldsymbol{\beta}; \mathbf{y})$, the model describes the data well if

$$L(\boldsymbol{\beta}; \mathbf{y}) \simeq L(\boldsymbol{\beta}_{\max}; \mathbf{y})$$

or poorly if

$$L(\boldsymbol{\beta}; \mathbf{y}) \ll L(\boldsymbol{\beta}_{\max}; \mathbf{y}).$$

This suggests the use of the generalized *likelihood ratio statistic* as a measure of goodness of fit,

$$\lambda = \frac{L(\mathbf{b}_{\max}; \mathbf{y})}{L(\mathbf{b}; \mathbf{y})}$$

or, equivalently,

$$\log \lambda = l(\mathbf{b}_{\max}; \mathbf{y}) - l(\mathbf{b}; \mathbf{y})$$

where $l(\mathbf{b}; \mathbf{y})$ is the log-likelihood function evaluated at the maximum likelihood estimator $\mathbf{b}$. Large values of $\log \lambda$ provide evidence that $\boldsymbol{\beta}$ is a poor model for the data. To determine the critical region for $\log \lambda$ we need to know the sampling distribution of $l(\mathbf{b}; \mathbf{y})$.

5.6 Sampling distribution for the log-likelihood function

The Taylor series approximation obtained by expanding $l(\boldsymbol{\beta};\mathbf{y})$ about the maximum likelihood estimator $\mathbf{b}$ is

$$l(\boldsymbol{\beta};\mathbf{y}) \simeq l(\mathbf{b};\mathbf{y})+(\boldsymbol{\beta}-\mathbf{b})^{\mathrm{T}}\mathbf{U}(\mathbf{b})+\tfrac{1}{2}(\boldsymbol{\beta}-\mathbf{b})^{\mathrm{T}}\mathbf{H}(\mathbf{b})(\boldsymbol{\beta}-\mathbf{b}) \qquad (5.5)$$

where $\mathbf{H}(\mathbf{b})$ is the matrix of second derivatives

$$\left[\frac{\partial^2 l}{\partial\beta_j\partial\beta_k}\right]$$

evaluated at $\mathbf{b}$. From the definition of $\mathbf{b}$, $\mathbf{U}(\mathbf{b}) = \mathbf{0}$. Also for large samples $\mathbf{H}(\mathbf{b})$ can be approximated using $\mathscr{I} = E[-\mathbf{H}]$. Thus (5.5) can be rewritten as

$$2[l(\mathbf{b};\mathbf{y})-l(\boldsymbol{\beta};\mathbf{y})] = (\boldsymbol{\beta}-\mathbf{b})^{\mathrm{T}}\mathscr{I}(\boldsymbol{\beta}-\mathbf{b}).$$

But $(\mathbf{b}-\boldsymbol{\beta})^{\mathrm{T}}\mathscr{I}(\mathbf{b}-\boldsymbol{\beta}) \sim \chi^2_p$ from result (5.3), so that

$$2[l(\mathbf{b};\mathbf{y})-l(\boldsymbol{\beta};\mathbf{y})] \sim \chi^2_p. \qquad (5.6)$$

We use a test statistic based on this result to assess the fit of a model and to compare alternative models.

5.7 Log-likelihood ratio statistic

We define

$$D = 2\log\lambda = 2[l(\mathbf{b}_{\max};\mathbf{y})-l(\mathbf{b};\mathbf{y})]. \qquad (5.7)$$

Nelder and Wedderburn (1972) called this the (scaled) *deviance*. It can be rewritten as

$$D = 2\{[l(\mathbf{b}_{\max};\mathbf{y})-l(\boldsymbol{\beta}_{\max};\mathbf{y})]-[l(\mathbf{b};\mathbf{y})-l(\boldsymbol{\beta};\mathbf{y})]+[l(\boldsymbol{\beta}_{\max};\mathbf{y})-l(\boldsymbol{\beta};\mathbf{y})]\}. \qquad (5.8)$$

The first expression on the right-hand side of (5.8) has the χ^2_N distribution (by result (5.6)); similarly the second has the χ^2_p distribution, and the third is a constant which is positive but near zero if the model based on $\boldsymbol{\beta}$ describes the data nearly as well as the maximal model does. Thus

$$D \sim \chi^2_{N-p}$$

if the model is good. If the model is poor the third term on the right-hand side of (5.8) will be large and so D will be larger than predicted by χ^2_{N-p} (in fact D has the non-central chi-squared distribution in this case).

Example 5.3

Suppose that response variables $Y_1, \ldots, Y_N$ are independent and Normally distributed with common variance σ^2. The log-likelihood function is

$$l(\boldsymbol{\beta}; \mathbf{y}) = -\frac{1}{2\sigma^2} \sum_{i=1}^{N} [y_i - E(Y_i)]^2 - \tfrac{1}{2} N \log(2\pi\sigma^2)$$

For the maximal model $E(Y_i) = \mu_i$, $i = 1, \ldots, N$ so $\boldsymbol{\beta}_{\max} = [\mu_1, \ldots, \mu_N]^T$ and $\hat{\mu}_i = y_i$. Therefore

$$l(\mathbf{b}_{\max}; \mathbf{y}) = -\tfrac{1}{2} N \log(2\pi\sigma^2)$$

For the model $E(Y_i) = \mu$ for all i, $\boldsymbol{\beta} = [\mu]^T$ and $\hat{\mu} = \bar{y}$. Therefore

$$l(\mathbf{b}; \mathbf{y}) = -\frac{1}{2\sigma^2} \sum_{i=1}^{N} (y_i - \bar{y})^2 - \tfrac{1}{2} N \log(2\pi\sigma^2)$$

Thus from definition (5.7)

$$D = \frac{1}{\sigma^2} \sum_{i=1}^{N} (y_i - \bar{y})^2.$$

If the model $Y_i \sim \text{NID}(\mu, \sigma^2)$ for all i is correct then $D \sim \chi^2_{N-1}$. Otherwise, for example if the Y_is do not all have the same mean, D will be larger than expected from the χ^2_{N-1} distribution.

5.8 Hypothesis testing

Consider the null hypothesis

$$H_0: \boldsymbol{\beta} = [\beta_1, \ldots, \beta_q]^T = \boldsymbol{\beta}_0$$

and a more general hypothesis

$$H_1: \boldsymbol{\beta} = [\beta_1, \ldots, \beta_p]^T = \boldsymbol{\beta}_1 \qquad \text{where } q < p < N.$$

We test H_0 against H_1 using the difference of log-likelihood ratio statistics

$$\begin{aligned} D = D_0 - D_1 &= 2[l(\mathbf{b}_{\max}; \mathbf{y}) - l(\mathbf{b}_0; \mathbf{y})] - 2[l(\mathbf{b}_{\max}; \mathbf{y}) - l(\mathbf{b}_1; \mathbf{y})] \\ &= 2[l(\mathbf{b}_1; \mathbf{y}) - l(\mathbf{b}_0; \mathbf{y})]. \end{aligned}$$

If both models describe the data well then $D_0 \sim \chi^2_{N-q}$ and $D_1 \sim \chi^2_{N-p}$ so that $D \sim \chi^2_{p-q}$ and we would generally prefer the model corresponding to H_0 because it is more parsimonious.

If the observed value of D is in the critical region (i.e. greater than the upper tail $100\,\alpha\,\%$ point of the χ^2_{p-q} distribution) then we would reject H_0 in favour of H_1 on the grounds that $\boldsymbol{\beta}_1$ provides a significantly better description of the data (even though it too may not fit the data particularly well).

Example 5.4

Consider two random samples $Y_{11}, \ldots, Y_{1K}$ and $Y_{21}, \ldots, Y_{2K}$ with $Y_{jk} \sim N(\mu_j, \sigma^2)$; for instance, the plant growth example (2.1). We want to test the null hypothesis $H_0: \mu_1 = \mu_2$ against the more general hypothesis $H_1: \mu_1$ and μ_2 are not necessarily equal. The models corresponding to H_0 and H_1 are summarized in Table 5.2.

Table 5.2 Summary of the hypothesis test for the equality of two means based on equal sized samples.

	Hypothesis	
	H_0	H_1
Model	$\boldsymbol{\beta}_0 = [\mu]$	$\boldsymbol{\beta}_1 = [\mu_1, \mu_2]^T$
Estimates	$\hat{\mu} = \bar{y} = \frac{1}{2K}\sum_j \sum_k y_{jk}$	$\hat{\mu}_j = \bar{y}_j = \frac{1}{K}\sum_k y_{jk}, \quad j = 1, 2$
Log-likelihood ratio statistic	$D_0 = \frac{1}{\sigma^2}\sum_j \sum_k (y_{jk} - \bar{y})^2$	$D_1 = \frac{1}{\sigma^2}\sum_j \sum_k (y_{jk} - \bar{y}_j)^2$
Sampling distribution	$D_0 \sim \chi^2_{2K-1}$ if H_0 is correct	$D_1 \sim \chi^2_{2K-2}$ if H_1 is correct

It can easily be shown that $D_0 - D_1 = (K/2\sigma^2)(\bar{y}_1 - \bar{y}_2)^2$. But we cannot use this to test H_0 directly if σ^2 is unknown. We estimate σ^2 by the pooled sample variance

$$s^2 = \frac{\Sigma\Sigma(y_{jk} - \bar{y}_j)^2}{2K-2} = \frac{\sigma^2 D_1}{2K-2}.$$

If H_1 is correct then $D_1 \sim \chi^2_{2K-2}$. If H_0 is also correct then $D_0 \sim \chi^2_{2K-1}$ so $D_0 - D_1 \sim \chi^2_1$ and hence

$$f = \frac{D_0 - D_1}{1} \bigg/ \frac{D_1}{2K-2} \sim F_{1,2K-2}.$$

If H_0 is not correct D_0 has a non-central chi-squared distribution so f has a non-central F distribution. Therefore values of the statistic f which are large compared with the $F_{1,2K-2}$ distribution provide evidence against H_0. Also

$$f = \frac{D_0 - D_1}{D_1/(2K-2)} = \frac{K(\bar{y}_1 - \bar{y}_2)^2}{2s^2} = \left[\frac{\bar{y}_1 - \bar{y}_2}{s\left(\frac{1}{K} + \frac{1}{K}\right)^{\frac{1}{2}}}\right]^2 = T^2,$$

which shows that this test is equivalent to the usual t-test for the equality of means for two independent samples (see Example 2.1).

5.9 Exercises

5.1 For the data in Example 5.2,

(i) calculate the log-likelihood ratio statistic D_1 for the Poisson model with $E(Y_i) = \beta_1 + \beta_2 x_i$,

(ii) fit the Poisson model with $E(Y_i) = \beta$ and compare the log-likelihood ratio statistic D_0 with D_1 from (i) to test the hypothesis that $E(Y_i)$ does not depend on x_i.

5.2 Find the log-likelihood ratio statistics for the following distributions. In each case consider a random sample $Y_1, \ldots, Y_N$ and the model $E(Y_i) = \mu$ for all i.

(i) Binomial distribution: $f(y;\pi) = \binom{n}{y}\pi^y(1-\pi)^{n-y}$

(ii) Poisson distribution: $f(y;\lambda) = \dfrac{\lambda^y e^{-\lambda}}{y!}$

(iii) Gamma distribution: $f(y;\theta) = \dfrac{y^{\phi-1}\theta^{\phi}e^{-y\theta}}{\Gamma(\phi)}$

where ϕ is a nuisance parameter.

(Nelder and Wedderburn, 1972)

5.3 Suppose that the link function $g(\mu)$ belongs to a family of link functions defined in terms of say two extra parameters $g(\mu;\alpha,\delta)$. For some application let $g_c(\mu;\alpha_c,\delta_c)$ denote the correct link function (i.e. $g_c(\mu;\alpha_c,\delta_c) = \mathbf{x}_i^T\boldsymbol{\beta}$) and let $g_0(\mu;\alpha_0,\delta_0)$ denote the link function actually used. Use a Taylor series expansion of g_c about α_0 and δ_0 to obtain the approximation

$$g_c(\mu;\alpha_0,\delta_0) = \mathbf{x}_i^T\boldsymbol{\beta} + \mathbf{d}^T\boldsymbol{\gamma}$$

where

$$\mathbf{d} = \left[\frac{\partial g_0}{\partial \alpha}, \frac{\partial g_0}{\partial \delta}\right]^T$$

evaluated at $\alpha = \alpha_0$ and $\delta = \delta_0$ and $\boldsymbol{\gamma} = [\alpha_0 - \alpha_c,\ \delta_0 - \delta_c]^T$. Hence suggest a method for testing the adequacy of the choice of link function.

(Pregibon, 1980)

6
MULTIPLE REGRESSION

6.1 Introduction

We begin the discussion of particular generalized linear models by considering the simplest case,

$$\mathbf{y} = \mathbf{X}\boldsymbol{\beta} + \mathbf{e} \tag{6.1}$$

where

(i) $\mathbf{y}$ is an $N \times 1$ response vector;
(ii) $\mathbf{X}$ is an $N \times p$ matrix of constants, with linearly independent columns so that rank $(\mathbf{X}) = p$;
(iii) $\boldsymbol{\beta}$ is a $p \times 1$ vector of parameters;
(iv) $\mathbf{e}$ is an $N \times 1$ random vector whose elements are independent, identically and Normally distributed $e_i \sim \mathrm{N}(0, \sigma^2)$ for $i = 1, \ldots, N$.

Since $E(\mathbf{y}) = \mathbf{X}\boldsymbol{\beta}$ the link function is the identity $g(\mu) = \mu$.

First we consider several examples of regression. Then we review the theoretical results relating to regression models; most of these have been obtained already as examples and exercises in previous chapters. Finally we mention several practical aspects of the use of regression. More detailed discussion of regression analyses can be found, for example, in the books by Draper and Smith (1981), Neter and Wasserman (1974), Seber (1977) or Graybill (1976).

6.2 Examples

Example 6.1 *Simple linear regression*

A straight-line relationship between a continuous variable, which is assumed to be Normally distributed, and a single explanatory variables is modelled by

$$E(Y_i) = \beta_0 + \beta_1 x_i, \qquad i = 1, \ldots, N.$$

This corresponds to the model $E(\mathbf{y}) = \mathbf{X}\boldsymbol{\beta}$ with

$$\mathbf{y} = \begin{bmatrix} Y_1 \\ \vdots \\ Y_N \end{bmatrix}, \quad \mathbf{X} = \begin{bmatrix} 1 & x_1 \\ \vdots & \vdots \\ 1 & x_N \end{bmatrix} \quad \text{and} \quad \boldsymbol{\beta} = \begin{bmatrix} \beta_0 \\ \beta_1 \end{bmatrix}.$$

The birth weight example in Section 2.3 involved models of this kind.

Example 6.2 *Multiple linear regression*

The data in Table 6.1 show responses, percentages of total calories obtained from complex carbohydrates, for 20 male insulin-dependent diabetics who had been on a high carbohydrate diet for 6 months. Compliance with the regime is thought to be related to age (in years), body weight (relative to 'ideal' weight for height) and other components of the diet, such as percentage of calories as protein. These other variables are treated as explanatory variables.

Table 6.1 Carbohydrate, age, weight and protein for 20 male insulin-dependent diabetics; for units, see text (data from K. Webb, personal communication).

Carbohydrate Y	Age x_1	Weight x_2	Protein x_3
33	33	100	14
40	47	92	15
37	49	135	18
27	35	144	12
30	46	140	15
43	52	101	15
34	62	95	14
48	23	101	17
30	32	98	15
38	42	105	14
50	31	108	17
51	61	85	19
30	63	130	19
36	40	127	20
41	50	109	15
42	64	107	16
46	56	117	18
24	61	100	13
35	48	118	18
37	28	102	14

If the response is linearly related to each of the covariates a suitable model is $E(\mathbf{y}) = \mathbf{X}\boldsymbol{\beta}$ with

$$\mathbf{y} = \begin{bmatrix} Y_1 \\ \vdots \\ Y_N \end{bmatrix}, \quad \mathbf{X} = \begin{bmatrix} 1 & x_{11} & x_{12} & x_{13} \\ \vdots & \vdots & \vdots & \vdots \\ 1 & x_{N1} & x_{N2} & x_{N3} \end{bmatrix} \quad \text{and} \quad \boldsymbol{\beta} = \begin{bmatrix} \beta_0 \\ \vdots \\ \beta_3 \end{bmatrix}$$

where $N = 20$. We use these data for illustrative purposes later in this chapter.

Example 6.3 *Polynomial regression*

A curvilinear relationship between the response variable Y and a single explanatory variable x may be modelled by a polynomial,

$$Y_i = \beta_0 + \beta_1 x_i + \beta_2 x_i^2 + \ldots + \beta_{p-1} x_i^{p-1} + e_i. \tag{6.2}$$

This is a special case of model (6.1) with

$$\mathbf{y} = \begin{bmatrix} Y_1 \\ \vdots \\ Y_N \end{bmatrix}, \quad \mathbf{X} = \begin{bmatrix} 1 & x_1 & x_1^2 \ldots x_1^{p-1} \\ \vdots & \vdots & \vdots \quad\quad \vdots \\ 1 & x_N & x_N^2 \quad\; x_N^{p-1} \end{bmatrix} \quad \text{and} \quad \boldsymbol{\beta} = \begin{bmatrix} \beta_0 \\ \vdots \\ \beta_{p-1} \end{bmatrix}$$

so that the powers of x_i are treated as distinct covariates. In practice it is inadvisable to use more than three or four terms in model (6.2) for several reasons:

(i) The columns of $\mathbf{X}$ are closely related and if p is too large $\mathbf{X}^T\mathbf{X}$ may be nearly singular (see Section 6.9);

(ii) There is a danger of producing a model which fits the data very well within the range of observations but is poor for prediction outside this range;

(iii) Often it is implausible that the mechanism linking x and Y is really described by a high order polynomial and an alternative formulation should be sought.

Example 6.4 *Trigonometric regression*

If the relationship between the response Y and an explanatory variable x is cyclic or *periodic* a suitable model might be

$$Y_i = \beta_0 + \beta_1 \cos \alpha_1 x_i + \beta_2 \sin \alpha_1 x_i + \beta_3 \cos \alpha_2 x_i + \beta_4 \sin \alpha_2 x_i + \ldots + e_i \tag{6.3}$$

where the α_js are known constants. In this case $E(\mathbf{y}) = \mathbf{X}\boldsymbol{\beta}$ with

$$\mathbf{X} = \begin{bmatrix} 1 & \cos \alpha_1 x_1 & \sin \alpha_1 x_1 & \cos \alpha_2 x_1 & \sin \alpha_2 x_1 & \ldots \\ \vdots & \vdots & \vdots & \vdots & \vdots & \\ 1 & \cos \alpha_1 x_N & \sin \alpha_1 x_N & \cos \alpha_2 x_N & \sin \alpha_2 x_N & \ldots \end{bmatrix}.$$

Trigonometric regression can be used to model seasonality in economic data, circadian rhythms and other periodic biological phenomena. For the same reasons as mentioned for polynomial regression, usually it is inadvisable to have too many terms on the right hand side of (6.3).

6.3 Maximum likelihood estimation

For independent random variables $Y_i \sim N(\mathbf{x}_i^T \boldsymbol{\beta}, \sigma^2)$, $i = 1, \ldots, N$ the log-likelihood function is

$$l = \frac{-1}{2\sigma^2}(\mathbf{y}-\mathbf{X}\boldsymbol{\beta})^T(\mathbf{y}-\mathbf{X}\boldsymbol{\beta}) - \tfrac{1}{2}N\log(2\pi\sigma^2) \tag{6.4}$$

where
$$\mathbf{y} = \begin{bmatrix} Y_1 \\ \vdots \\ Y_N \end{bmatrix}, \quad \mathbf{X} = \begin{bmatrix} \mathbf{x}_1^T \\ \vdots \\ \mathbf{x}_N^T \end{bmatrix} \quad \text{and} \quad \boldsymbol{\beta} = \begin{bmatrix} \beta_1 \\ \vdots \\ \beta_p \end{bmatrix}.$$

We assume that rank $(\mathbf{X}) = p$ so that the $p \times p$ matrix $\mathbf{X}^T\mathbf{X}$ is non-singular. From (6.4)

$$\mathbf{U} = \frac{\partial l}{\partial \boldsymbol{\beta}} = \frac{1}{\sigma^2}\mathbf{X}^T(\mathbf{y}-\mathbf{X}\boldsymbol{\beta})$$

so the maximum likelihood estimator of $\boldsymbol{\beta}$ is given by the solution of $\mathbf{X}^T\mathbf{X}\mathbf{b} = \mathbf{X}^T\mathbf{y}$, i.e.

$$\mathbf{b} = (\mathbf{X}^T\mathbf{X})^{-1}\mathbf{X}^T\mathbf{y}. \tag{6.5}$$

In Example 5.1 it is shown that $E(\mathbf{b}) = \boldsymbol{\beta}$ and $E[(\mathbf{b}-\boldsymbol{\beta})(\mathbf{b}-\boldsymbol{\beta})^T] = \sigma^2(\mathbf{X}^T\mathbf{X})^{-1}$. Also $\mathbf{b}$ is a linear combination of elements of the Normally distributed responses Y_i so that

$$\mathbf{b} \sim N(\boldsymbol{\beta}, \sigma^2(\mathbf{X}^T\mathbf{X})^{-1}). \tag{6.6}$$

(More generally if $Y_i \sim N(\mathbf{x}_i^T\boldsymbol{\beta}, \sigma_i^2)$ then the maximum likelihood estimator of $\boldsymbol{\beta}$ is the solution of $\mathbf{X}^T\mathbf{V}^{-1}\mathbf{X}\mathbf{b} = \mathbf{X}^T\mathbf{V}^{-1}\mathbf{y}$ where $\mathbf{V}$ is the diagonal matrix with elements $v_{ii} = \sigma_i^2$ – see Exercise 4.3.)

For generalized linear models σ^2 is treated as a constant and it is not estimated so the distribution (6.6) is not fully determined. More conventionally $\boldsymbol{\beta}$ and σ^2 are estimated simultaneously to give

$$\mathbf{b} = (\mathbf{X}^T\mathbf{X})^{-1}\mathbf{X}^T\mathbf{y} \quad \text{and} \quad \tilde{\sigma}^2 = \frac{1}{N}(\mathbf{y}-\mathbf{X}\mathbf{b})^T(\mathbf{y}-\mathbf{X}\mathbf{b}).$$

However, it can be shown that $E(\tilde{\sigma}^2) = (N-p)\sigma^2/N$ so that

$$\hat{\sigma}^2 = \frac{1}{N-p}(\mathbf{y}-\mathbf{X}\mathbf{b})^T(\mathbf{y}-\mathbf{X}\mathbf{b}) \tag{6.7}$$

provides an unbiased estimator of σ^2. Using results (6.6) and (6.7) confidence intervals and hypothesis tests for $\boldsymbol{\beta}$ can be derived.

6.4 Least squares estimation

If $E(\mathbf{y}) = \mathbf{X}\boldsymbol{\beta}$ and $E[(\mathbf{y}-\mathbf{X}\boldsymbol{\beta})(\mathbf{y}-\mathbf{X}\boldsymbol{\beta})^T] = \mathbf{V}$ where $\mathbf{V}$ is known we can obtain the least squares estimator of $\boldsymbol{\beta}$ without making any further assumptions about the distribution of $\mathbf{y}$. We minimize

$$S_w = (\mathbf{y}-\mathbf{X}\boldsymbol{\beta})^T \mathbf{V}^{-1} (\mathbf{y}-\mathbf{X}\boldsymbol{\beta}).$$

The solution of

$$\frac{\partial S_w}{\partial \boldsymbol{\beta}} = -2\mathbf{X}^T \mathbf{V}^{-1} (\mathbf{y}-\mathbf{X}\boldsymbol{\beta}) = 0$$

is

$$\mathbf{b} = (\mathbf{X}^T \mathbf{V}^{-1} \mathbf{X})^{-1} \mathbf{X}^T \mathbf{V}^{-1} \mathbf{y}$$

(provided that the matrix inverses exist) – see Section 4.3. In particular if the elements of $\mathbf{y}$ are independent and have a common variance then

$$\mathbf{b} = (\mathbf{X}^T\mathbf{X})^{-1}\mathbf{X}^T\mathbf{y}.$$

Thus for regression models with Normal errors, maximum likelihood and least squares estimators are the same.

6.5 Log-likelihood ratio statistic

For a maximal model in which $\boldsymbol{\beta}_{\max} = [\beta_1, \ldots, \beta_N]^T$, without loss of generality we can take $\mathbf{X}$ as the identity matrix so that $\mathbf{b}_{\max} = \mathbf{y}$ and by substitution in (6.4)

$$l(\mathbf{b}_{\max}; \mathbf{y}) = -\tfrac{1}{2} N \log(2\pi\sigma^2).$$

For any other model $E(\mathbf{y}) = \mathbf{X}\boldsymbol{\beta}$ involving p parameters $\boldsymbol{\beta} = [\beta_1, \ldots, \beta_p]^T$ where $p < N$, let $\mathbf{b}$ denote the maximum likelihood estimator. Then the log-likelihood ratio statistic is

$$\begin{aligned} D &= 2[l(\mathbf{b}_{\max}; \mathbf{y}) - l(\mathbf{b}; \mathbf{y})] \\ &= \frac{1}{\sigma^2}(\mathbf{y}-\mathbf{X}\mathbf{b})^T(\mathbf{y}-\mathbf{X}\mathbf{b}) \\ &= \frac{1}{\sigma^2}(\mathbf{y}^T\mathbf{y} - 2\mathbf{b}^T\mathbf{X}^T\mathbf{y} + \mathbf{b}^T\mathbf{X}^T\mathbf{X}\mathbf{b}) \\ &= \frac{1}{\sigma^2}(\mathbf{y}^T\mathbf{y} - \mathbf{b}^T\mathbf{X}^T\mathbf{y}). \end{aligned}$$

because $\mathbf{X}^T\mathbf{X}\mathbf{b} = \mathbf{X}^T\mathbf{y}$.

If the model is correct then $D \sim \chi^2_{N-p}$, otherwise D has the non-central chi-squared distribution with $N-p$ degrees of freedom (from Section 5.7).

The statistic D is not completely determined when σ^2 is unknown. As illustrated by the numerical examples in Chapter 2, for hypothesis testing we overcome this difficulty by using appropriately defined ratios of log-likelihood ratio statistics as follows.

Consider two models

$$E(\mathbf{y}) = \mathbf{X}_0\boldsymbol{\beta}_0 \tag{6.8}$$

and

$$E(\mathbf{y}) = \mathbf{X}_1\boldsymbol{\beta}_1 \tag{6.9}$$

with $\boldsymbol{\beta}_0 = [\beta_1, ..., \beta_q]^{\mathrm{T}}$ and $\boldsymbol{\beta}_1 = [\beta_1, ..., \beta_p]^{\mathrm{T}}$ where $q < p < N$ such that model (6.8) is a special case of model (6.9); model (6.8) is sometimes called the *reduced model*. These models correspond to hypotheses H_0 and H_1, respectively, where H_0 is a 'null' hypothesis and H_1 is not a conventional alternative hypothesis but instead is a more general hypothesis than H_0. Let $\mathbf{b}_0$ and $\mathbf{b}_1$ be the maximum likelihood estimators of $\boldsymbol{\beta}_0$ and $\boldsymbol{\beta}_1$ and let D_0 and D_1 denote the corresponding statistics. The difference in fit between the models is

$$D_0 - D_1 = \frac{1}{\sigma^2}[(\mathbf{y}^{\mathrm{T}}\mathbf{y} - \mathbf{b}_0^{\mathrm{T}}\mathbf{X}_0^{\mathrm{T}}\mathbf{y}) - (\mathbf{y}^{\mathrm{T}}\mathbf{y} - \mathbf{b}_1^{\mathrm{T}}\mathbf{X}_1^{\mathrm{T}}\mathbf{y})] = \frac{1}{\sigma^2}(\mathbf{b}_1^{\mathrm{T}}\mathbf{X}_1^{\mathrm{T}}\mathbf{y} - \mathbf{b}_0^{\mathrm{T}}\mathbf{X}_0^{\mathrm{T}}\mathbf{y}).$$

We assume that the more general model (6.9) is correct so that $D_1 \sim \chi^2_{N-p}$. If model (6.8) is also correct (i.e. if H_0 is true) then $D_0 \sim \chi^2_{N-q}$, otherwise D_0 has a non-central chi-squared distribution with $N-q$ degrees of freedom. Thus if H_0 is true

$$D_0 - D_1 \sim \chi^2_{p-q}$$

so

$$f = \frac{D_0 - D_1}{p-q} \bigg/ \frac{D_1}{N-p} = \frac{\mathbf{b}_1^{\mathrm{T}}\mathbf{X}_1^{\mathrm{T}}\mathbf{y} - \mathbf{b}_0^{\mathrm{T}}\mathbf{X}_0^{\mathrm{T}}\mathbf{y}}{p-q} \bigg/ \frac{\mathbf{y}^{\mathrm{T}}\mathbf{y} - \mathbf{b}_1^{\mathrm{T}}\mathbf{X}_1^{\mathrm{T}}\mathbf{y}}{N-p} \sim F_{p-q,N-p}.$$

If H_0 is not true, f has a non-central F distribution. Therefore, values of f which are large relative to $F_{p-q,N-p}$ provide evidence against H_0. This test for H_0 is often set out as shown in Table 6.2.

Table 6.2 Analysis of variance table.

Source of variation	Degrees of freedom	Sum of squares	Mean square
Model (6.8)	q	$\mathbf{b}_0^{\mathrm{T}}\mathbf{X}_0^{\mathrm{T}}\mathbf{y}$	
Improvement due to model (6.9)	$p-q$	$\mathbf{b}_1^{\mathrm{T}}\mathbf{X}_1^{\mathrm{T}}\mathbf{y} - \mathbf{b}_0^{\mathrm{T}}\mathbf{X}_0^{\mathrm{T}}\mathbf{y}$	$\dfrac{\mathbf{b}_1^{\mathrm{T}}\mathbf{X}_1^{\mathrm{T}}\mathbf{y} - \mathbf{b}_0^{\mathrm{T}}\mathbf{X}_0^{\mathrm{T}}\mathbf{y}}{p-q}$
Residual	$N-p$	$\mathbf{y}^{\mathrm{T}}\mathbf{y} - \mathbf{b}_1^{\mathrm{T}}\mathbf{X}_1^{\mathrm{T}}\mathbf{y}$	$\dfrac{\mathbf{y}^{\mathrm{T}}\mathbf{y} - \mathbf{b}_1^{\mathrm{T}}\mathbf{X}_1^{\mathrm{T}}\mathbf{y}}{N-p}$
Total	N	$\mathbf{y}^{\mathrm{T}}\mathbf{y}$	

This form of analysis is one of the major tools for inference from regression models (and also for analysis of variance – see Chapter 7). It depends on the assumption that the most general model fitted, in this case model (6.9), does describe the data well so that the corresponding statistic D has the central chi-squared distribution. This assumption should, if possible, be checked; for example by examining the residuals (see Section 6.8).

Another less rigorous comparison of goodness of fit between two models is provided by R^2, the square of the multiple correlation coefficient.

6.6 Multiple correlation coefficient and R^2

If $\mathbf{y} = \mathbf{X}\boldsymbol{\beta}+\mathbf{e}$ and the elements of $\mathbf{e}$ are independent with $E(e_i) = 0$ and $\text{var}(e_i) = \sigma^2$ for $i = 1, \ldots, N$, then the least squares criterion is

$$S = \sum_{i=1}^{N} e_i^2 = \mathbf{e}^T\mathbf{e} = (\mathbf{y}-\mathbf{X}\boldsymbol{\beta})^T(\mathbf{y}-\mathbf{X}\boldsymbol{\beta}).$$

The minimum value of S for the model is

$$\hat{S} = (\mathbf{y}-\mathbf{X}\mathbf{b})^T(\mathbf{y}-\mathbf{X}\mathbf{b}) = \mathbf{y}^T\mathbf{y}-\mathbf{b}^T\mathbf{X}^T\mathbf{y}$$

(from Section 6.4). This can be used as a measure of the fit of the model.

The value of $\hat{S}$ is compared with the fit of the simplest or *minimal model* $E(y_i) = \mu$ for all i. This model can be written in the general form $E(\mathbf{y}) = \mathbf{X}\boldsymbol{\beta}$ if $\boldsymbol{\beta} = [\mu]$ and $\mathbf{X} = \mathbf{1}$, the $N \times 1$ vector of ones. Therefore $\mathbf{X}^T\mathbf{X} = N$, $\mathbf{X}^T\mathbf{y} = \Sigma y_i$ and $\mathbf{b} = \hat{\mu} = \bar{y}$. The corresponding value of the least squares criterion is

$$S_0 = \mathbf{y}^T\mathbf{y}-N\bar{y}^2 = \sum_{i=1}^{N} (y_i-\bar{y})^2.$$

Thus S_0 is proportional to the variance of the observations and it is regarded as the 'worst' possible value of S.

Any other model with value $\hat{S}$ is assessed relative to S_0. The difference

$$S_0-\hat{S} = \mathbf{b}^T\mathbf{X}^T\mathbf{y}-N\bar{y}^2$$

is the improvement in fit due to the model $E(\mathbf{y}) = \mathbf{X}\boldsymbol{\beta}$.

$$R^2 = \frac{S_0-\hat{S}}{S_0} = \frac{\mathbf{b}^T\mathbf{X}^T\mathbf{y}-N\bar{y}^2}{\mathbf{y}^T\mathbf{y}-N\bar{y}^2}$$

is interpreted as the proportion of the total variation in the data which is explained by the model.

If the model does not describe the data any better than the minimal model then $S_0 \simeq \hat{S}$ so $R^2 \simeq 0$. If the maximal model with N parameters is used then $\mathbf{b} = \mathbf{y}$ and $\mathbf{b}^T\mathbf{X}^T\mathbf{y} = \mathbf{y}^T\mathbf{y}$ so that $R^2 = 1$ corresponding to a 'perfect' fit. In general $0 < R^2 < 1$. The square root of R^2 is called the *multiple correlation coefficient*.

A disadvantage of using R^2 as a measure of goodness of fit is that its sampling distribution is not readily determined. Also its value is not adjusted for the number of parameters used in the fitted model.

The use of R^2 and hypothesis testing based on the log-likelihood ratio statistic are illustrated in the following numerical example.

6.7 Numerical example

We use the carbohydrate data in Table 6.1 and begin by fitting the model

$$E(Y_i) = \beta_0 + \beta_1 x_{i1} + \beta_2 x_{i2} + \beta_3 x_{i3} \tag{6.10}$$

in which carbohydrate Y is linearly related to age x_1, weight x_2 and protein x_3 $(i = 1, \dots, N = 20)$. If

$$\mathbf{X} = \begin{bmatrix} 1 & x_{11} & x_{12} & x_{13} \\ \vdots & \vdots & \vdots & \vdots \\ 1 & x_{N1} & x_{N2} & x_{N3} \end{bmatrix} \quad \text{and} \quad \boldsymbol{\beta} = \begin{bmatrix} \beta_0 \\ \vdots \\ \beta_3 \end{bmatrix}$$

then

$$\mathbf{X}^{\mathrm{T}}\mathbf{y} = \begin{bmatrix} 752 \\ 34596 \\ 82270 \\ 12105 \end{bmatrix} \quad \text{and} \quad \mathbf{X}^{\mathrm{T}}\mathbf{X} = \begin{bmatrix} 20 & 923 & 2214 & 318 \\ 923 & 45697 & 102003 & 14780 \\ 2214 & 102003 & 250346 & 35306 \\ 318 & 14780 & 35306 & 5150 \end{bmatrix}.$$

The solution of $\mathbf{X}^{\mathrm{T}}\mathbf{X}\mathbf{b} = \mathbf{X}^{\mathrm{T}}\mathbf{y}$ is

$$\mathbf{b} = \begin{bmatrix} 36.9601 \\ -0.1137 \\ -0.2280 \\ 1.9577 \end{bmatrix} \quad \text{and} \quad (\mathbf{X}^{\mathrm{T}}\mathbf{X})^{-1} = \begin{bmatrix} 4.8158 & -0.0113 & -0.0188 & -0.1362 \\ -0.0113 & 0.0003 & 0.0000 & -0.0004 \\ -0.0188 & 0.0000 & 0.0002 & -0.0002 \\ -0.1362 & -0.0004 & -0.0002 & 0.0114 \end{bmatrix}$$

correct to 4 decimal places. Also $\mathbf{y}^{\mathrm{T}}\mathbf{y} = 29368$, $N\bar{y}^2 = 28275.2$ and $\mathbf{b}^{\mathrm{T}}\mathbf{X}^{\mathrm{T}}\mathbf{y} = 28800.337$ so that $R^2 = 0.48$, i.e. 48% of the total variation in the data is explained by model (6.10). Using (6.7), for an unbiased estimator of σ^2 we obtain $\hat{\sigma}^2 = 35.479$ and the standard errors for elements of $\mathbf{b}$ shown in Table 6.3.

Table 6.3 Estimates for model (6.10).

Term	Estimate b_j	Standard error*
Constant	36.960	13.071
Coefficient for age	−0.114	0.109
Coefficient for weight	−0.228	0.084
Coefficient for protein	1.958	0.635

* Values calculated using more significant figures for $(\mathbf{X}^{\mathrm{T}}\mathbf{X})^{-1}$ than shown above.

To illustrate the use of the log-likelihood ratio statistic we test the hypothesis that the response does not depend on age, i.e. $H_0: \beta_1 = 0$. The corresponding model is

$$E(Y_i) = \beta_0 + \beta_2 x_{i2} + \beta_3 x_{i3}. \tag{6.11}$$

The matrix **X** for this model is obtained from the previous one by omitting the second column so that

$$\mathbf{X}^T\mathbf{y} = \begin{bmatrix} 752 \\ 82270 \\ 12105 \end{bmatrix}, \qquad \mathbf{X}^T\mathbf{X} = \begin{bmatrix} 20 & 2214 & 318 \\ 2214 & 250346 & 35306 \\ 318 & 35306 & 5150 \end{bmatrix}$$

and hence

$$\mathbf{b} = \begin{bmatrix} 33.130 \\ -0.222 \\ 1.824 \end{bmatrix}.$$

For model (6.11) $\mathbf{b}^T\mathbf{X}^T\mathbf{y} = 28761.978$ so that $R^2 = 0.445$, i.e. $44\frac{1}{2}\%$ of the variation is explained by the model. The significance test for H_0 is summarized in Table 6.4.

Table 6.4 Analysis of variance table comparing models (6.10) and (6.11).

Source of variation	Degrees of freedom	Sum of squares	Mean square
Model (6.11)	3	28761.978	
Improvement due to model (6.10)	1	38.359	38.36
Residual	16	567.663	35.48
Total	20	29368.000	

$f = 38.36/35.48 = 1.08$ which is not significant compared with the $F_{1,16}$ distribution so the data provide no evidence against H_0, i.e. the response appears to be unrelated to age.

6.8 Residuals

For the regression model (6.1) we assume that the error terms $e_i = Y_i - E(Y_i)$ are independent, identically distributed with $e_i \sim N(0, \sigma^2)$ for all i and that they do not vary in magnitude with elements of **y** or **X**. Also we assume that the most extensive model that we fit describes the data well so that the corresponding log-likelihood ratio statistic has a central chi-squared distribution.

It is important to check these assumptions by examining the residuals

$$\hat{\mathbf{e}} = \mathbf{y} - \mathbf{X}\mathbf{b}.$$

If the model is correct then $E(\hat{\mathbf{e}}) = \mathbf{0}$ and

$$E(\hat{\mathbf{e}}\hat{\mathbf{e}}^{\mathrm{T}}) = E(\mathbf{y}\mathbf{y}^{\mathrm{T}}) - \mathbf{X}\,E(\mathbf{b}\mathbf{b}^{\mathrm{T}})\mathbf{X}^{\mathrm{T}} = \sigma^2[\mathbf{I} - \mathbf{X}(\mathbf{X}^{\mathrm{T}}\mathbf{X})^{-1}\mathbf{X}^{\mathrm{T}}] \tag{6.12}$$

where $\mathbf{I}$ is the unit matrix. So the *standardized residuals* are defined by

$$r_i = \frac{\hat{e}_i}{\hat{\sigma}(1-p_{ii})^{\frac{1}{2}}}$$

where p_{ii} is the ith element on the diagonal of the matrix $\mathbf{P} = \mathbf{X}(\mathbf{X}^{\mathrm{T}}\mathbf{X})^{-1}\mathbf{X}^{\mathrm{T}}$.

The frequency distribution of the r_is (or the $\hat{e}_i$s) should be plotted to give a rough check for Normality and to identify unusual values. A more accurate test of Normality is obtained by plotting the $\hat{r}_i$s against the expected Normal order statistics and seeing if the points lie on a straight line. As shown by (6.12), the residuals are slightly correlated, however substantial serial correlations between them may indicate mis-specification of the model, so it is often worthwhile to check for serial correlation (for example, using the Durbin–Watson test).

Also the residuals should be plotted against the predicted values $\hat{\mathbf{y}} = \mathbf{X}\mathbf{b}$ and against each of the explanatory variables. Patterns in these plots also indicate mis-specifications of the model (see Exercise 6.2). In addition they can be used to identify any unusual observations which may have a strong influence on the value of $\mathbf{b}$ and on the goodness of fit of the model.

An excellent discussion of the examination of residuals is given in Chapter 3 of the book by Draper and Smith (1981).

6.9 Orthogonality

In the numerical example in Section 6.7 the parameters β_0, β_2 and β_3 occurred in both models (6.10) and (6.11) but their estimates differed when the models were fitted to the data. Also the test of the hypothesis that $\beta_1 = 0$ would depend on which terms were included in the model. For example, the analysis of variance table comparing the models

$$E(Y_i) = \beta_0 + \beta_1 x_{i1} + \beta_3 x_{i3} \tag{6.13}$$

and

$$E(Y_i) = \beta_0 + \beta_3 x_{i3} \tag{6.14}$$

(which do not include $\beta_2 x_{i2}$) will differ from Table 6.4 comparing models (6.10) and (6.11) – see Exercise 6.1 (iii).

Usually estimates, confidence intervals and hypothesis tests depend on which covariates, other than the ones of direct interest, are included in the model. An exception is when the $N \times p$ matrix $\mathbf{X}$ is orthogonal, i.e. it can be partitioned into components

$$\mathbf{X} = [\mathbf{X}_1, \ldots, \mathbf{X}_m], \qquad m \leqslant p,$$

such that $\mathbf{X}_j^{\mathrm{T}}\mathbf{X}_k = \mathbf{0}$ where the matrices $\mathbf{X}_j$ correspond to submodels of interest. Let $\boldsymbol{\beta} = [\boldsymbol{\beta}_1, \ldots, \boldsymbol{\beta}_m]^{\mathrm{T}}$ be the corresponding partition of the parameters, then

$$E(\mathbf{y}) = \mathbf{X}\boldsymbol{\beta} = \mathbf{X}_1\boldsymbol{\beta}_1 + \ldots + \mathbf{X}_m\boldsymbol{\beta}_m$$

and $\mathbf{X}^{\mathrm{T}}\mathbf{X}$ is the block diagonal matrix

$$\mathbf{X}^{\mathrm{T}}\mathbf{X} = \begin{bmatrix} \mathbf{X}_1^{\mathrm{T}}\mathbf{X}_1 & & \mathbf{O} \\ & \ddots & \\ \mathbf{O} & & \mathbf{X}_m^{\mathrm{T}}\mathbf{X}_m \end{bmatrix}, \qquad \text{also } \mathbf{X}^{\mathrm{T}}\mathbf{y} = \begin{bmatrix} \mathbf{X}_1^{\mathrm{T}}\mathbf{y} \\ \vdots \\ \mathbf{X}_m^{\mathrm{T}}\mathbf{y} \end{bmatrix}.$$

where $\mathbf{O}$ is used to denote that the remaining elements of the matrix are zeros. Therefore the estimates $\mathbf{b}_j = (\mathbf{X}_j^{\mathrm{T}}\mathbf{X}_j)^{-1}\mathbf{X}_j^{\mathrm{T}}\mathbf{y}$ are unaltered by the omission or inclusion of other components in the model and

$$\mathbf{b}^{\mathrm{T}}\mathbf{X}^{\mathrm{T}}\mathbf{y} = \mathbf{b}_1^{\mathrm{T}}\mathbf{X}_1^{\mathrm{T}}\mathbf{y} + \ldots + \mathbf{b}_m^{\mathrm{T}}\mathbf{X}_m^{\mathrm{T}}\mathbf{y}.$$

Also the hypotheses

$$\mathrm{H}_1: \boldsymbol{\beta}_1 = \mathbf{0}, \ldots, \quad \mathrm{H}_m: \boldsymbol{\beta}_m = \mathbf{0}$$

can be tested independently as shown in Table 6.5.

Table 6.5 Multiple hypothesis tests when **X** is orthogonal

Source of variation	Degrees of freedom	Sum of squares
Model corresponding to H_1	p_1	$\mathbf{b}_1^{\mathrm{T}}\mathbf{X}_1^{\mathrm{T}}\mathbf{y}$
$\vdots$	$\vdots$	$\vdots$
Model corresponding to H_m	p_m	$\mathbf{b}_m^{\mathrm{T}}\mathbf{X}_m^{\mathrm{T}}\mathbf{y}$
Residual	$N - \sum_{j=1}^{m} p_j$	$\mathbf{y}^{\mathrm{T}}\mathbf{y} - \mathbf{b}^{\mathrm{T}}\mathbf{X}^{\mathrm{T}}\mathbf{y}$
Total	N	$\mathbf{y}^{\mathrm{T}}\mathbf{y}$

Unfortunately the benefits of orthogonality can only be exploited if **X** can be designed to have this property. This may be possible if the elements of **X** are dummy variables representing factor levels (see Chapter 7) or if polynomial regression is performed using *orthogonal polynomials* (these are specially constructed polynomials such that the columns of **X** corresponding to successively higher powers of the explanatory variable are orthogonal – see Draper and Smith, 1981, Section 5.6).

6.10 Collinearity

If the explanatory variables are closely related to one another the columns of $\mathbf{X}$ may nearly be linearly dependent so that $\mathbf{X}^T\mathbf{X}$ is nearly singular. In this case the equation $\mathbf{X}^T\mathbf{X}\mathbf{b} = \mathbf{X}^T\mathbf{y}$ is said to be *ill-conditioned* and the solution $\mathbf{b}$ will be unstable in the sense that small changes in the data may cause large changes in $\mathbf{b}$. Also at least some of the elements of $(\mathbf{X}^T\mathbf{X})^{-1}$ will be large corresponding to large variance and covariance estimates of $\mathbf{b}$. Thus careful inspection of the matrix $\hat{\sigma}^2(\mathbf{X}^T\mathbf{X})^{-1}$ may reveal the presence of collinearity.

The resolution of the problem is more difficult. It may require extra information from the substantive area from which the data came, an alternative specification of the model or some other non-computational approach. In addition various computational techniques, such as *ridge regression*, have been proposed for handling this problem. Detailed discussions of collinearity are given, for example, in the books by Belsey *et al.* (1980) and Draper and Smith (1981).

A particular difficulty with collinearity occurs in the selection of some subset of the explanatory variables which 'best' describes the data. If two variables are highly correlated it may be impossible, on statistical grounds alone, to determine which should be included in a model.

6.11 Model selection

Many applications of regression involve large numbers of explanatory variables and an important issue is to identify a subset of these variables which provides a good and parsimonious model for the response. The usual procedure is sequentially to add or delete terms from the model; this is called *stepwise regression*. Unless the variables are orthogonal this involves considerable computation and multiple testing of related hypotheses (with associated difficulties in interpreting significance levels).

For further consideration of these problems the reader is referred to any of the standard textbooks on regression (also see Exercise 6.2 and the discussion of collinearity in Section 6.10).

6.12 Non-linear regression

The term *non-linear regression* is used for two types of models.

First, $E(Y) = g(\mathbf{x}^T\boldsymbol{\beta})$ – which is a generalized linear model if the distribution of Y is a member of the exponential family. An example is Holliday's (1960) equation for plant yield

$$E(Y) = \frac{1}{\beta_0 + \beta_1 x + \beta_2 x^2}$$

where Y is the yield per plant and x is a measure of plant density. If we assume that Y is Normally distributed then we can use the methods of Chapters 4 and 5 for estimation and inference.

Second, $E(Y) = g(\mathbf{x}, \boldsymbol{\beta})$ where g is intrinsically non-linear in the parameters, for example, the logistic growth model

$$E(Y) = \frac{\beta_0}{1+\beta_1 e^{\beta_2 x}}.$$

For these cases iterative estimation procedures can be derived by methods analogous to those used in Chapter 4 (Charnes *et al.* 1976; Ratkowsky and Dolby, 1975). However, the sampling distributions of the estimators may be seriously non-Normal (Gillis and Ratkowsky, 1978).

6.13 Exercises

6.1 Analyse the carbohydrate data in Table 6.1 using an appropriate computer program (or, preferably, repeat the analyses using several different regression programs and compare the results).

(i) Plot the responses $\mathbf{y}$ against each of the explanatory variables $\mathbf{x}_1$, $\mathbf{x}_2$ and $\mathbf{x}_3$ to see if $\mathbf{y}$ appears to be linearly related to them.

(iii) Fit the full model (6.10) and examine the residuals using the methods suggested in Section 6.8.

(iii) Fit models (6.13) and (6.14) and use these to test the hypothesis: $\beta_1 = 0$. Compare your results with Table 6.4.

6.2 Suppose that the true relationship between $\mathbf{y}$ and blocks of covariates $\mathbf{X}_1$ and $\mathbf{X}_2$ is

$$E(\mathbf{y}) = \mathbf{X}_1\boldsymbol{\beta}_1 + \mathbf{X}_2\boldsymbol{\beta}_2$$

but you postulate the model

$$E(\mathbf{y}) = \mathbf{X}_1\boldsymbol{\beta}_1. \qquad (6.15)$$

Show that the estimator $\mathbf{b}_1$ from the mis-specified model (6.15) is biased.

(Draper and Smith, 1981, Section 2.12)

7
ANALYSIS OF VARIANCE AND COVARIANCE

7.1 Introduction

This chapter concerns linear models of the form

$$\mathbf{y} = \mathbf{X}\boldsymbol{\beta} + \mathbf{e} \quad \text{with} \quad \mathbf{e} \sim \mathrm{N}(\mathbf{0}, \sigma^2\mathbf{I})$$

where $\mathbf{y}$ and $\mathbf{e}$ are random vectors of length N, $\mathbf{X}$ is an $N \times p$ matrix of constants, $\boldsymbol{\beta}$ is a vector of p parameters and $\mathbf{I}$ is the unit matrix. These models differ from the regression models of the previous chapter in that $\mathbf{X}$, called the *design matrix*, consists entirely of dummy variables for analysis of variance (ANOVA) or dummy variables and measured covariates for analysis of covariance (ANCOVA). Since the choice of dummy variables is to some extent arbitrary, a major consideration is the optimal choice of $\mathbf{X}$. The main questions addressed by analysis of variance and covariance involve comparisons of means. Traditionally the emphasis is on hypothesis testing rather than estimation or prediction.

In this book we only consider *fixed effects models* in which the levels of factors are regarded as fixed so that $\boldsymbol{\beta}$ is a vector of constants. We do not consider *random effects models* where the factor levels are regarded as a random selection from a population of possible levels and $\boldsymbol{\beta}$ is treated as a vector of random variables. The problem of estimating variances for the elements of $\boldsymbol{\beta}$ in random effects models, also called *variance components* models, is discussed by McCullagh and Nelder (1983) in the framework of generalized linear models. Also the elements of the response vector $\mathbf{y}$ are assumed to be independent and therefore we do not consider situations involving *repeated measures* on the same experimental units so that the observations are likely to be correlated.

Wider coverage of analysis of variance and covariance is provided by any of the conventional books on the subject, for example Graybill (1976), Searle (1971), Scheffé (1959) or Winer (1971).

7.2 Basic results

Since the random components $\mathbf{e}$ in ANOVA and ANCOVA models are assumed to be Normally distributed many of the results obtained in the previous chapter apply here too. For instance the log-likelihood function is

$$l = -\frac{1}{2\sigma^2}(\mathbf{y}-\mathbf{X}\boldsymbol{\beta})^{\mathrm{T}}(\mathbf{y}-\mathbf{X}\boldsymbol{\beta}) - \frac{N}{2}\log(2\pi\sigma^2)$$

so the maximum likelihood (or least squares) estimator $\mathbf{b}$ is the solution of the normal equations

$$\mathbf{X}^{\mathrm{T}}\mathbf{X}\mathbf{b} = \mathbf{X}^{\mathrm{T}}\mathbf{y}. \tag{7.1}$$

In ANOVA models there are usually more parameters than there are independent equations in $E(\mathbf{y}) = \mathbf{X}\boldsymbol{\beta}$ therefore $\mathbf{X}^{\mathrm{T}}\mathbf{X}$ is singular and there is no unique solution of (7.1). In this case $\boldsymbol{\beta}$ is said to not be *estimable* or *identifiable*. To obtain a particular solution extra equations are used so that $\mathbf{b}$ is the solution of

$$\left.\begin{aligned} \mathbf{X}^{\mathrm{T}}\mathbf{X}\mathbf{b} &= \mathbf{X}^{\mathrm{T}}\mathbf{y} \\ \text{and} \quad \mathbf{C}\mathbf{b} &= \mathbf{0} \end{aligned}\right\}. \tag{7.2}$$

In anticipation of the need for the extra equations $\mathbf{Cb} = \mathbf{0}$, the model $E(\mathbf{y}) = \mathbf{X}\boldsymbol{\beta}$ often includes the *constraint equations* $\mathbf{C}\boldsymbol{\beta} = \mathbf{0}$. However, the minimum of $(\mathbf{y}-\mathbf{X}\boldsymbol{\beta})^{\mathrm{T}}(\mathbf{y}-\mathbf{X}\boldsymbol{\beta})$ is unique and it is given by any solution of (7.1) (see Exercise 7.4), so the value of $(\mathbf{y}-\mathbf{Xb})^{\mathrm{T}}(\mathbf{y}-\mathbf{Xb})$ does not depend on the choice of the constraint equations. Other properties of $\mathbf{b}$ do depend on the choice of $\mathbf{C}$ as illustrated in the numerical examples in Sections 7.3 and 7.4.

For a maximal model $\boldsymbol{\beta}_{\max}$ has N elements $[\beta_1, \beta_2, \ldots, \beta_N]^{\mathrm{T}}$. Therefore, without loss of generality, we can take $\mathbf{X}$ to be the $N \times N$ unit matrix $\mathbf{I}$ so that $\mathbf{b}_{\max} = \mathbf{y}$ and hence

$$l(\mathbf{b}_{\max}; \mathbf{y}) = -\frac{N}{2}\log(2\pi\sigma^2).$$

For any other model with p parameters and estimator $\mathbf{b}$, the log-likelihood ratio statistic is

$$D = 2[l(\mathbf{b}_{\max}; \mathbf{y}) - l(\mathbf{b}; \mathbf{y})] = \frac{1}{\sigma^2}(\mathbf{y}-\mathbf{Xb})^{\mathrm{T}}(\mathbf{y}-\mathbf{Xb}) = \frac{1}{\sigma^2}(\mathbf{y}^{\mathrm{T}}\mathbf{y} - \mathbf{b}^{\mathrm{T}}\mathbf{X}^{\mathrm{T}}\mathbf{y}). \tag{7.3}$$

If the model is correct $D \sim \chi^2_{N-p}$, otherwise D has a non-central chi-squared distribution. As with regression models D is not completely determined when σ^2 is unknown so that hypotheses are tested by comparing appropriate ratios of log-likelihood ratio statistics and using the F-distribution.

7.3 One factor ANOVA

The data in Table 7.1 are an extension of the plant weight example of Chapter 2. An experiment is conducted to compare yields (as measured by dried weight of plants) obtained under a control and two different treatment conditions. Thus the response, plant weight, depends on one factor, growing condition, with three levels – control, treatment A and treatment B. We are interested in whether response means differ between the three groups.

Table 7.1 Plant weights from three different growing conditions.

Control	4.17	5.58	5.18	6.11	4.50	4.61	5.17	4.53	5.33	5.14
Treatment A	4.81	4.17	4.41	3.59	5.87	3.83	6.03	4.89	4.32	4.69
Treatment B	6.31	5.12	5.54	5.50	5.37	5.29	4.92	6.15	5.80	5.26

More generally, if experimental units are randomly allocated to groups corresponding to J levels of a factor, this is called a *completely randomized experimental design* and the data can be set out as in Table 7.2.

Table 7.2 Data for one factor ANOVA with J levels of the factor and unequal sample sizes.

Factor level	Responses			Totals
A_1	Y_{11}	Y_{12} ...	Y_{1n_1}	$Y_{1.}$
A_2	Y_{21}	Y_{22} ...	Y_{2n_2}	$Y_{2.}$
$\vdots$				
A_J	Y_{J1}	Y_{J2} ...	Y_{Jn_J}	$Y_{J.}$

The responses can be written as the vector

$$\mathbf{y} = [Y_{11}, \ldots, Y_{1n_1}, Y_{21}, \ldots, Y_{2n_2}, \ldots, Y_{Jn_J}]^{\mathrm{T}}$$

of length $N = \sum_{j=1}^{J} n_j$. For simplicity we only consider the case when all the samples are of the same size, i.e. $n_j = K$ for all j so $N = JK$.

We consider three different formulations of the model corresponding to the hypothesis that the response means differ for different levels of the factor. The simplest version of the model is

$$E(Y_{jk}) = \mu_j, \qquad j = 1, \ldots, J. \tag{7.4}$$

In terms of the vector $\mathbf{y}$ this can be written as

$$E(Y_i) = \sum_{j=1}^{J} x_{ij}\mu_j, \qquad i = 1, \ldots, N$$

where $x_{ij} = 1$ if response Y_i corresponds to level A_j and $x_{ij} = 0$ otherwise.

Thus $E(\mathbf{y}) = \mathbf{X}\boldsymbol{\beta}$

with
$$\boldsymbol{\beta} = \begin{bmatrix} \mu_1 \\ \mu_2 \\ \vdots \\ \mu_J \end{bmatrix} \quad \text{and} \quad \mathbf{X} = \begin{bmatrix} \mathbf{1} & \mathbf{0} \ldots \mathbf{0} \\ \mathbf{0} & \mathbf{1} & & \mathbf{O} \\ \vdots & & \ddots & \\ \mathbf{0} & \mathbf{O} & & \mathbf{1} \end{bmatrix}$$

where
$$\mathbf{0} = \begin{bmatrix} 0 \\ 0 \\ \vdots \\ 0 \end{bmatrix} \quad \text{and} \quad \mathbf{1} = \begin{bmatrix} 1 \\ 1 \\ \vdots \\ 1 \end{bmatrix}$$

are vectors of length K. Then $\mathbf{X}^T\mathbf{X}$ is the $J \times J$ diagonal matrix

$$\mathbf{X}^T\mathbf{X} = \begin{bmatrix} K & 0 \ldots 0 \\ 0 & K & & \mathbf{O} \\ \vdots & & \ddots & \\ 0 & \mathbf{O} & & K \end{bmatrix} \quad \text{and} \quad \mathbf{X}^T\mathbf{y} = \begin{bmatrix} Y_{1.} \\ Y_{2.} \\ \vdots \\ Y_{J.} \end{bmatrix} \quad \text{so that } \mathbf{b} = \frac{1}{K}\begin{bmatrix} Y_{1.} \\ Y_{2.} \\ \vdots \\ Y_{J.} \end{bmatrix} = \begin{bmatrix} \bar{y}_1 \\ \bar{y}_2 \\ \vdots \\ \bar{y}_J \end{bmatrix}$$

and
$$\mathbf{b}^T\mathbf{X}^T\mathbf{y} = \frac{1}{K}\sum_{j=1}^{J} Y_{j.}^2.$$

In addition, the fitted values are $\hat{\mathbf{y}} = [\bar{y}_1, \bar{y}_1, \ldots, \bar{y}_1, \bar{y}_2, \ldots, \bar{y}_J]^T$. The disadvantage of this simple formulation of the model is that it cannot be extended to more than one factor. For generalizability, we need to specify the model so that parameters for levels and combinations of levels of factors reflect differential effects beyond some average response.

One such formulation is

$$E(Y_{jk}) = \mu + \alpha_j, \qquad j = 1, \ldots, J,$$

where μ is the average effect and α_j is an additional effect due to A_j. For this parameterization there are $J+1$ parameters.

$$\boldsymbol{\beta} = \begin{bmatrix} \mu \\ \alpha_1 \\ \vdots \\ \alpha_J \end{bmatrix}, \qquad \mathbf{X} = \begin{bmatrix} \mathbf{1} & \mathbf{1} & \mathbf{0} \ldots \mathbf{0} \\ \mathbf{1} & \mathbf{0} & \mathbf{1} & & \mathbf{O} \\ \vdots & & & \ddots & \\ \mathbf{1} & & \mathbf{O} & & \mathbf{1} \end{bmatrix}$$

where $\mathbf{0}$ and $\mathbf{1}$ are vectors of length K. Thus

$$\mathbf{X}^T\mathbf{y} = \begin{bmatrix} Y_{..} \\ Y_{1.} \\ \vdots \\ Y_{J.} \end{bmatrix} \quad \text{and} \quad \mathbf{X}^T\mathbf{X} = \begin{bmatrix} N & K \ldots K \\ K & K & & \mathbf{O} \\ \vdots & & \ddots & \\ K & \mathbf{O} & & K \end{bmatrix}.$$

The first row of the $(J+1) \times (J+1)$ matrix $\mathbf{X}^T\mathbf{X}$ is the sum of the remaining rows so $\mathbf{X}^T\mathbf{X}$ is singular and there is no unique solution of the normal equations $\mathbf{X}^T\mathbf{X}\mathbf{b} = \mathbf{X}^T\mathbf{y}$. The general solution can be written as

$$\mathbf{b} = \begin{bmatrix} \hat{\mu} \\ \hat{\alpha}_1 \\ \vdots \\ \hat{\alpha}_J \end{bmatrix} = \frac{1}{K}\begin{bmatrix} 0 \\ Y_{1.} \\ \vdots \\ Y_{J.} \end{bmatrix} - \lambda \begin{bmatrix} -1 \\ 1 \\ \vdots \\ 1 \end{bmatrix}$$

where λ is an arbitrary constant. It is traditional to impose the additional '*sum to zero*' constraint

$$\sum_{j=1}^{J} \hat{\alpha}_j = 0 \qquad \text{so that } \frac{Y_{..}}{K} = J\lambda, \quad \text{i.e. } \lambda = \frac{Y_{..}}{N} \text{ since } N = JK,$$

giving the solution

$$\hat{\mu} = \frac{Y_{..}}{N} \quad \text{and} \quad \hat{\alpha}_j = \frac{Y_{j.}}{K} - \frac{Y_{..}}{N} \qquad \text{for } j = 1, \ldots, J.$$

Hence

$$\mathbf{b}^T\mathbf{X}^T\mathbf{y} = \frac{Y_{..}^2}{N} - \sum_{j=1}^{J} Y_{j.}\left(\frac{Y_{j.}}{K} - \frac{Y_{..}}{N}\right) = \frac{1}{K}\sum_{j=1}^{J} Y_{j.}^2$$

and the fitted values are $\hat{\mathbf{y}} = [\bar{y}_1, \bar{y}_1, \ldots, \bar{y}_J]^T$, as for the first version of the model.

A third version of the model is $E(Y_{jk}) = \mu + \alpha_j$ with the constraint that $\alpha_1 = 0$ so that α_j measures the difference between the first and jth levels of the factor and μ represents the effect of the first level. This is called a *corner-point parameterization*. For this version there are J parameters.

$$\boldsymbol{\beta} = \begin{bmatrix} \mu \\ \alpha_2 \\ \vdots \\ \alpha_J \end{bmatrix}, \qquad \text{also} \qquad \mathbf{X} = \begin{bmatrix} \mathbf{1} & \mathbf{0} & \ldots & \mathbf{0} \\ \mathbf{1} & \mathbf{1} & & \mathbf{O} \\ \vdots & & \ddots & \\ \mathbf{1} & \mathbf{O} & & \mathbf{1} \end{bmatrix}$$

so

$$\mathbf{X}^T\mathbf{y} = \begin{bmatrix} Y_{..} \\ Y_{2.} \\ \vdots \\ Y_{J.} \end{bmatrix} \qquad \text{and} \qquad \mathbf{X}^T\mathbf{X} = \begin{bmatrix} N & K & \ldots & K \\ K & K & & \mathbf{O} \\ \vdots & & \ddots & \\ K & \mathbf{O} & & K \end{bmatrix}.$$

The $J \times J$ matrix $\mathbf{X}^T\mathbf{X}$ is non-singular so there is a unique solution

$$\mathbf{b} = \frac{1}{K}\begin{bmatrix} & Y_{1.} & \\ Y_{2.} & - & Y_{1.} \\ & \vdots & \\ Y_{J.} & - & Y_{1.} \end{bmatrix}$$

for the normal (7.1). Hence

$$\mathbf{b}^T\mathbf{X}^T\mathbf{y} = \frac{1}{K}\left[Y_{..}Y_{1.} + \sum_{j=2}^{J} Y_{j.}(Y_{j.} - Y_{1.})\right] = \frac{1}{K}\sum_{j=1}^{J} Y_{j.}^2$$

and the fitted values $\hat{\mathbf{y}} = [\bar{y}_1, \bar{y}_1, \ldots, \bar{y}_J]^T$ as before.

Table 7.3 ANOVA table for one factor with J levels and equal sample size K per level.

Source of variation	Degrees of freedom	Sum of squares	Mean square	f
Mean	1	$\frac{1}{N}Y_{..}^2$		
Between levels of factor A	$J-1$	$\frac{1}{K}\sum_{j=1}^{J} Y_{j.}^2 - \frac{1}{N}Y_{..}^2$ $= \sigma^2(D_0-D_1)$	$\frac{\sigma^2(D_0-D_1)}{J-1}$	$\frac{D_0-D_1}{J-1}\Big/\frac{D_1}{N-J}$
Residual	$N-J$	$\sum_{j=1}^{J}\sum_{k=1}^{J} Y_{jk}^2 - \frac{1}{K}\sum_{j=1}^{J} Y_{j.}^2$ $= \sigma^2 D_1$	$\sigma^2 D_1/N-J$	
Total	N	$\sum_{j=1}^{J}\sum_{k=1}^{K} Y_{jk}^2$		

Thus although the three specifications of the model differ, the values of $\hat{\mathbf{y}} = \mathbf{Xb}$, $\mathbf{b}^T\mathbf{X}^T\mathbf{y}$ and

$$D_1 = \frac{1}{\sigma^2}(\mathbf{y}^T\mathbf{y} - \mathbf{b}^T\mathbf{X}^T\mathbf{y}) = \frac{1}{\sigma^2}\left[\sum_{j=1}^{J}\sum_{k=1}^{K} Y_{jk}^2 - \frac{1}{K}\sum_{j=1}^{J} Y_{j.}^2\right]$$

are the same in each case.

These three versions of the model $E(Y_{jk}) = \mu_j$ all correspond to the hypothesis H_1: the response means for each level may differ. To compare this with the null hypothesis H_0: the means are all equal, we consider the model $E(Y_{jk}) = \mu$ so that $\boldsymbol{\beta} = [\mu]$ and $\mathbf{X} = [\mathbf{1}\,\mathbf{1} \ldots \mathbf{1}]^T$. Then $\mathbf{X}^T\mathbf{X} = N$, $\mathbf{X}^T\mathbf{y} = Y_{..}$ and hence $\mathbf{b} = \hat{\mu} = Y_{..}/N$ with $\mathbf{b}^T\mathbf{X}^T\mathbf{y} = Y_{..}^2/N$ and

$$D_0 = \frac{1}{\sigma^2}\left[\sum_{j=1}^{J}\sum_{k=1}^{K} Y_{jk}^2 - \frac{Y_{..}^2}{N}\right].$$

To test H_0 against H_1 we assume that H_1 is correct so that $D_1 \sim \chi^2_{N-J}$. If, in addition, H_0 is correct then $D_0 \sim \chi^2_{N-1}$, otherwise D_0 has a non-central chi-squared distribution. Thus if H_0 is correct

$$D_0 - D_1 = \frac{1}{\sigma^2}\left[\frac{1}{K}\sum_{j=1}^{J} Y_{j.}^2 - \frac{1}{N}Y_{..}^2\right] \sim \chi^2_{J-1}$$

and so

$$f = \frac{D_0-D_1}{J-1}\Big/\frac{D_1}{N-J} \sim F_{J-1,N-J};$$

if H_0 is not correct f is likely to be larger than predicted from the $F_{J-1,N-J}$ distribution. Conventionally this hypothesis test is set out as in Table 7.3.

For the plant weight data the results are summarized in Table 7.4.

Table 7.4 ANOVA table for plant weight data in Table 7.1.

Source of variation	Degrees of freedom	Sum of squares	Mean square	f
Mean	1	772.060		
Between treatments	2	3.766	1.833	4.85
Residual	27	10.492	0.389	
Total	30	786.318		

Since $f = 4.85$ is significant at the 5% level when compared with the $F_{2,27}$ distribution, we conclude that the mean responses differ. From the model $E(Y_{jk}) = \mu_j$ the means are

$$\mathbf{b} = \begin{bmatrix} \hat{\mu}_1 \\ \hat{\mu}_2 \\ \hat{\mu}_3 \end{bmatrix} = \begin{bmatrix} 5.032 \\ 4.661 \\ 5.526 \end{bmatrix}.$$

We assume that H_1 is correct so that $D_1 \sim \chi^2_{27}$. Using the observed value of D_1, $10.492/\sigma^2$, and the fact that $E(\chi^2_n) = n$ we estimate σ^2 by

$$\hat{\sigma}^2 = \frac{10.492}{E(D_1)} = \frac{10.492}{27} = 0.389$$

(i.e. the residual mean square in Table 7.4). Thus the standard error of each mean is $(0.389/10)^{\frac{1}{2}} = 0.197$ because the variance–covariance matrix of $\mathbf{b}$ is $\sigma^2(\mathbf{X}^T\mathbf{X})^{-1}$ where

$$\mathbf{X}^T\mathbf{X} = \begin{bmatrix} 10 & 0 & 0 \\ 0 & 10 & 0 \\ 0 & 0 & 10 \end{bmatrix}.$$

Now it can be seen that the significant effect is due to the mean for treatment B being significantly larger than the others.

7.4 Two factor ANOVA with replication

Consider the fictitious data in Table 7.5 in which factor A (with $J = 3$ levels) and factor B (with $K = 2$ levels) are *crossed* so that there are JK subclasses formed by all combinations of A and B levels. In each subclass there are $L = 2$ observations or *replications*.

Table 7.5 Fictitious data for two factor ANOVA with equal numbers of observations in each subclass.

Levels of factor A	Levels of factor B		Total
	B_1	B_2	
A_1	6.8, 6.6	5.3, 6.1	24.8
A_2	7.5, 7.4	7.2, 6.5	28.6
A_3	7.8, 9.1	8.8, 9.1	34.8
Total	45.2	43.0	88.2

The main hypotheses are:

H_I: there are no interaction effects, i.e. the effects of A and B are additive;
H_A: there are no differences in response associated with different levels of factor A;
H_B: there are no differences in response associated with different levels of factor B.

Thus we need to consider a *full model* and three *reduced models* formed by omitting various terms from the full model.

(i) The full model is

$$E(Y_{jkl}) = \mu + \alpha_j + \beta_k + (\alpha\beta)_{jk} \tag{7.5}$$

where the terms $(\alpha\beta)_{jk}$ correspond to *interaction effects* and α_j and β_k to *main effects* of the factors.

(ii) The *additive model* is

$$E(Y_{jkl}) = \mu + \alpha_j + \beta_k. \tag{7.6}$$

This is compared to the full model to test hypothesis H_I.

(iii) The model formed by omitting effects due to B is

$$E(Y_{jkl}) = \mu + \alpha_j. \tag{7.7}$$

This is compared to the additive model to test hypothesis H_B.

(iv) The model formed by omitting effects due to A is

$$E(Y_{jkl}) = \mu + \beta_j. \tag{7.8}$$

This is compared to the additive model to test hypothesis H_A.

The models (7.5)–(7.8) have too many parameters; for instance replicates in the same subclass have the same expected value so there can be at most JK independent expected values but the full model has

$1+J+K+JK = (J+1)(K+1)$ parameters. To overcome this difficulty (which leads to the singularity of $\mathbf{X}^T\mathbf{X}$) we can impose the extra constraints

$$\alpha_1+\alpha_2+\alpha_3 = 0, \quad \beta_1+\beta_2 = 0,$$
$$(\alpha\beta)_{11}+(\alpha\beta)_{12} = 0, \quad (\alpha\beta)_{21}+(\alpha\beta)_{22} = 0, \quad (\alpha\beta)_{31}+(\alpha\beta)_{32} = 0,$$
$$(\alpha\beta)_{11}+(\alpha\beta)_{21}+(\alpha\beta)_{31} = 0$$

(the remaining condition $(\alpha\beta)_{12}+(\alpha\beta)_{22}+(\alpha\beta)_{32} = 0$ follows from the last four equations). These are the conventional constraint equations for ANOVA. Alternatively we can take $\alpha_1 = \beta_1 = (\alpha\beta)_{11} = (\alpha\beta)_{12} = (\alpha\beta)_{21} = (\alpha\beta)_{31} = 0$ as the corner-point constraints. In either case the numbers of (linearly) independent parameters are: 1 for μ, $J-1$ for the α_js, $K-1$ for the β_ks, and $(J-1)(K-1)$ for the $(\alpha\beta)_{jk}$s.

Details of fitting all four models using either the sum-to-zero constraints or the corner-point constraints are given in Appendix 3.

In general, hypothesis tests may not be statistically independent so that the order in which the models are fitted affects the results. For the data in Table 7.5, however, it is possible to specify the design matrix $\mathbf{X}$ so that it has orthogonal components corresponding to the mean, H_I, H_A and H_B and therefore the hypothesis tests are independent. Details of this orthogonal parameterization are also given in Appendix 3.

For models (7.5)–(7.8) the estimates $\mathbf{b}$ depend on the choice of constraints and dummy variables. However, the fitted values $\hat{\mathbf{y}} = \mathbf{Xb}$ are the same for all specifications of the models and so the values of $\mathbf{b}^T\mathbf{X}^T\mathbf{y}$ and $\sigma^2 D = \mathbf{y}^T\mathbf{y}-\mathbf{b}^T\mathbf{X}^T\mathbf{y}$ are the same. For these data $\mathbf{y}^T\mathbf{y} = 664.1$ and the other results are summarized in Table 7.6 (the subscripts F, I, A and B refer to the full model and the models corresponding to H_I, H_A and H_B, respectively).

Table 7.6 Summary of calculations for data in Table 7.5.

Terms in model	Hypothesis	Number of parameters	$\mathbf{b}^T\mathbf{X}^T\mathbf{y}$	$\sigma^2 D = \mathbf{y}^T\mathbf{y}-\mathbf{b}^T\mathbf{X}^T\mathbf{y}$
$\mu+\alpha_j+\beta_k+(\alpha\beta)_{jk}$		6	662.6200	$\sigma^2 D_F = 1.4800$
$\mu+\alpha_j+\beta_k$	H_I	4	661.4133	$\sigma^2 D_I = 2.6867$
$\mu+\alpha_j$	H_B	3	661.0100	$\sigma^2 D_B = 3.0900$
$\mu+\beta_k$	H_A	2	648.6733	$\sigma^2 D_A = 15.4267$
μ		1	648.2700	

To test H_I we assume that the full model is correct so that $D_F \sim \chi^2_6$ because there are $N = 12$ observations and the model has $JK = 6$ independent parameters. If H_I is correct also then $D_I \sim \chi^2_8$ so that $D_I - D_F \sim \chi^2_2$ and

$$f = \frac{D_I - D_F}{2} \Big/ \frac{D_F}{6} \sim F_{2,6}.$$

The value of

$$f = \frac{2.6867-1.48}{2\sigma^2} \bigg/ \frac{1.48}{6\sigma^2} = 2.45$$

is not significant so the data provide no evidence against H_I. Since H_I is not rejected we proceed to test H_A and H_B. For H_B we consider the difference in fit between the models $E(Y_{jkl}) = \mu + \alpha_j$ and $E(Y_{jkl}) = \mu + \alpha_j + \beta_k$ i.e. $D_B - D_I$ and compare this with D_F using

$$f = \frac{D_B - D_I}{1} \bigg/ \frac{D_F}{6} = \frac{3.09-2.6867}{\sigma^2} \bigg/ \frac{1.48}{6\sigma^2} = 1.63$$

which is not significant compared to the $F_{1,6}$ distribution, suggesting that there are no differences due to levels of factor B. The corresponding test for H_A gives $f = 25.82$ which is significant compared with the $F_{2,6}$ distribution. Thus we conclude that the response means are only affected by differences in the levels of factor A. These results are usually summarized as shown in Table 7.7.

Table 7.7 ANOVA table for data in Table 7.5.

Source of variation	Degrees of freedom	Sum of squares	Mean square	f
Mean	1	648.2700		
Levels of A	2	12.7400	6.3700	25.82
Levels of B	1	0.4033	0.4033	1.63
Interactions	2	1.2067	0.6033	2.45
Residual	6	1.4800	0.2467	
Total	12	664.1		

7.5 Crossed and nested factors and more complicated models

In the example in Section 7.4 the factors A and B are said to be *crossed* because there is a subclass corresponding to each combination of levels A_j and B_k and all the comparisons represented by the terms α_j, β_k and $(\alpha\beta)_{jk}$ in the full model $E(Y_{jkl}) = \mu + \alpha_j + \beta_k + (\alpha\beta)_{jk}$, $j = 1, \ldots, J$, $k = 1, \ldots, K$ are of potential interest.

This contrasts with the two factor *nested* design shown in Table 7.8 which represents an experiment to compare two drugs (A_1 and A_2) one of which is tested in three hospitals (B_1, B_2 and B_3) and the other in two hospitals (B_4 and B_5).

We want to compare the effects of the two drugs and possible differences in response between hospitals using the same drug. It is not sensible to make

Table 7.8 Nested two factor experiment.

	Drug A_1			Drug A_2	
Hospitals	B_1	B_2	B_3	B_4	B_5
Responses	Y_{111}	Y_{121}	Y_{131}	Y_{241}	Y_{251}
	$\vdots$	$\vdots$	$\vdots$	$\vdots$	$\vdots$
	Y_{11n_1}	Y_{12n_2}	Y_{13n_3}	Y_{24n_4}	Y_{25n_5}

comparisons between hospitals using different drugs. A suitable full model is $E(\mathbf{y}) = \mathbf{X}\boldsymbol{\beta}$ where

$$\boldsymbol{\beta} = [\mu, \alpha_1, \alpha_2, (\alpha\beta)_{11}, (\alpha\beta)_{12}, (\alpha\beta)_{13}, (\alpha\beta)_{24}, (\alpha\beta)_{25}]^{\mathrm{T}}$$

and the response vector $\mathbf{y}$ has length $N = \sum_{k=1}^{5} n_k$. For the conventional ANOVA constraints we let $\alpha_1 + \alpha_2 = 0$, $(\alpha\beta)_{11} + (\alpha\beta)_{12} + (\alpha\beta)_{13} = 0$ and $(\alpha\beta)_{24} + (\alpha\beta)_{25} = 0$, or for the corner-point constraints we take $\alpha_1 = (\alpha\beta)_{11} = (\alpha\beta)_{24} = 0$. Reduced models to compare hospitals using the same drug are formed by omitting the terms $(\alpha\beta)_{1k}$, $k = 1, 2, 3$ and, separately, $(\alpha\beta)_{2k}$, $k = 4, 5$. The reduced model for the hypothesis of no difference between the drugs (but allowing for differences between hospitals) is $E(\mathbf{y}) = \mathbf{X}\boldsymbol{\beta}$ with $\boldsymbol{\beta} = [\mu, \beta_1, \beta_2, \beta_3, \beta_4, \beta_5]^{\mathrm{T}}$ where the β_ks correspond to hospitals and $\beta_1 + \beta_2 + \beta_3 = 0$ and $\beta_4 + \beta_5 = 0$ or $\beta_1 = \beta_4 = 0$.

ANOVA models can readily be defined for more than two factors. The factors may be crossed or nested or some mixture of these forms. The models can include higher order interaction terms such as $(\alpha\beta\gamma)_{jkl}$ as well as the first order interactions like $(\alpha\beta)_{jk}$ and the main effects. These extensions do not involve any fundamental differences from the examples already considered so they are not examined further in this book.

7.6 More complicated hypotheses

In all the above examples we only considered hypotheses in which certain parameters in the full model are omitted in the reduced models. For instance, in the plant weight example $E(Y_{jk}) = \mu + \alpha_j$ in the full model and $E(Y_{jk}) = \mu$ in the reduced model corresponding to the hypothesis that $\alpha_1 = \alpha_2 = \alpha_3 = 0$. Sometimes we are interested in testing more complicated hypotheses such as treatments A and B in the plant weight experiment being equally effective but different from the control, i.e. $\alpha_2 = \alpha_3$ but α_1 not necessarily the same. Such hypotheses can be readily accommodated in the model fitting approach by the appropriate choice of parameters and dummy variables, for example the hypothesis $\alpha_2 = \alpha_3$ is equivalent to fitting $E(Y_{1k}) = \beta_1$ and $E(Y_{2k}) = E(Y_{3k}) = \beta_2$.

7.7 Independence of hypothesis tests

In the two factor ANOVA example in Section 7.4 the tests of the three hypotheses H_I, H_A, H_B are statistically independent because there is an orthogonal form of the design matrix $\mathbf{X}$ for the full model so that $\mathbf{X}^T\mathbf{X}$ is block diagonal with blocks corresponding to the mean and the three hypotheses. Hence the total sum of squares can be partitioned into disjoint components corresponding to the mean, H_I, H_A, H_B and the residual. For two factor ANOVA such a partition is only possible if the numbers n_{jk} of observations in each subclass satisfy $n_{jk} = n_{j.}n_{.k}/n_{..}$ (see Winer, 1971, Section 5.23–8).

In general, multiple hypothesis tests are only independent if there is a design matrix with orthogonal components so that the total sum of squares can be partitioned into disjoint terms corresponding to the hypotheses. Usually this is only possible if the hypotheses are particularly simple (e.g. interaction and main effects are zero) and if the experimental design is *balanced* (i.e. there are equal numbers of observations in each subclass). If the hypotheses are not independent then care is needed in interpreting simultaneous significance tests.

7.8 Choice of constraint equations and dummy variables

The numerical examples also illustrate several major issues relating to the choice of constraint equations and dummy variables for ANOVA models.

ANOVA models are usually specified in terms of parameters which are readily interpretable as marginal effects due to factor levels and interactions. However, the models contain more parameters than there are independent normal equations. Therefore extra equations, traditionally in the form of sum-to-zero constraints are added. (If the design is unbalanced there is some controversy about the most appropriate choice of constraint equations.) In the framework of generalized linear models this means that the equations (7.2) to be solved are not the normal equations obtained by the methods of maximum likelihood or least squares. Therefore the standard computational procedures cannot be used. Also the terms of $\boldsymbol{\beta}$ are generally not identifiable, and unique unbiased point estimates and confidence intervals can only be obtained for certain linear combinations of parameters, called *estimable functions*. Nevertheless, if the main purpose of analysing the data is to test hypotheses, the use of sum-to-zero constraints is entirely appropriate and convenient provided that special purpose computer programs are used.

If the corner-point constraints are used the elements of $\boldsymbol{\beta}$ and the corresponding columns of $\mathbf{X}$ are arranged as $\boldsymbol{\beta} = [\boldsymbol{\beta}_1, \boldsymbol{\beta}_2]^T$ and $\mathbf{X} = [\mathbf{X}_1, \mathbf{X}_2]$ so that $\mathbf{X}_1^T\mathbf{X}_1$ is non-singular and $\boldsymbol{\beta}_2$ is set to $\mathbf{0}$. Thus

$$E(\mathbf{y}) = \mathbf{X}\boldsymbol{\beta} = \mathbf{X}_1\boldsymbol{\beta}_1.$$

Then the normal equations

$$\mathbf{X}_1^{\mathrm{T}}\mathbf{X}_1\mathbf{b}_1 = \mathbf{X}_1^{\mathrm{T}}\mathbf{y}$$

can be solved using standard multiple regression or generalized linear modelling programs and the estimators have various desirable properties (e.g. $\mathbf{b}_1$ is unbiased and has variance–covariance matrix $\sigma^2(\mathbf{X}_1^{\mathrm{T}}\mathbf{X}_1)^{-1}$). However, the interpretation of parameters subject to corner-point constraints is perhaps less straightforward than with sum-to-zero constraints. Also all the calculations usually have to be repeated for each new model fitted. In practice, estimation using corner-point constraints is performed so that parameters are estimated sequentially in such a way that the redundant corner-point parameters (which are said to be *aliased*) are systematically identified and set equal to zero (for example, this is the procedure used in GLIM).

In the two factor ANOVA example in Section 7.4, the most elegant analysis was obtained by choosing the dummy variables so that the design matrix $\mathbf{X}$ had orthogonal components corresponding to each of the hypotheses to be tested. For simple well-planned experiments where this form of analysis is possible there are computational benefits (e.g. parameter estimates are the same for all models) and advantages in interpretation (e.g. independence of the hypothesis tests). However, for unbalanced experimental designs or hypotheses involving more complicated contrasts, it is unlikely that orthogonal forms exist.

In summary, for any particular sequence of models the choice of constraints and dummy variables affects the computational procedures and the parameter estimates. However, it does not influence the results for hypothesis testing. The reason is that any solution $\mathbf{b}$ of the normal equations (7.1) corresponds to the unique minimum of $(\mathbf{y}-\mathbf{X}\boldsymbol{\beta})^{\mathrm{T}}\ (\mathbf{y}-\mathbf{X}\boldsymbol{\beta})$. Hence the statistics $\sigma^2 D = \mathbf{y}^{\mathrm{T}}\mathbf{y}-\mathbf{b}^{\mathrm{T}}\mathbf{X}^{\mathrm{T}}\mathbf{y}$ are the same regardless of the way the models are specified.

7.9 Analysis of covariance

This is the term used for mixed models in which some of the explanatory variables are dummy variables representing factor levels and others are continuous measurements, called covariates. As with ANOVA we are interested in comparing means for subclasses defined by factor levels but, recognizing that the covariates may also affect the responses, we compare the means after 'adjustment' for covariate effects.

A typical example is provided by the data in Table 7.9. The responses Y_{jk} are achievement scores, the levels of the factor represent three different training methods and the covariates x_{jk} are aptitude scores measured before training commenced. We want to compare the training methods, taking into account differences in initial aptitude between the three groups of subjects.

Table 7.9 Achievement scores (data from Winer, 1971, p. 766).

Training method	A_1		A_2		A_3	
	y	x	y	x	y	x
	6	3	8	4	6	3
	4	1	9	5	7	2
	5	3	7	5	7	2
	3	1	9	4	7	3
	4	2	8	3	8	4
	3	1	5	1	5	1
	6	4	7	2	7	4
Total	31	15	53	24	47	19
Sum of squares	147	41	413	96	321	59
Σxy	75		191		132	

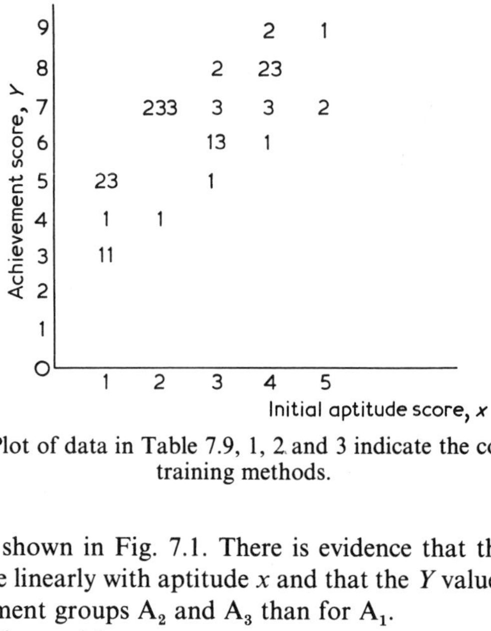

Figure 7.1 Plot of data in Table 7.9, 1, 2 and 3 indicate the corresponding training methods.

The data are shown in Fig. 7.1. There is evidence that the achievement scores Y increase linearly with aptitude x and that the Y values are generally higher for treatment groups A_2 and A_3 than for A_1.

We compare the models

$$E(Y_{jk}) = \mu_j + \gamma x_{jk} \tag{7.9}$$

and

$$E(Y_{jk}) = \mu + \gamma x_{jk} \tag{7.10}$$

for $j = 1, 2, 3$ and $k = 1, \ldots, 7$. Model (7.10) corresponds to the null hypothesis that there are no differences in mean achievement scores between the three training methods. Let $\mathbf{y}_j = [Y_{j1}, \ldots, Y_{j7}]^T$ and $\mathbf{x}_j = [x_{j1}, \ldots, x_{j7}]^T$ so that in matrix notation model (7.9) is $E(\mathbf{y}) = \mathbf{X}\boldsymbol{\beta}$ with

$$\mathbf{y} = \begin{bmatrix} \mathbf{y}_1 \\ \mathbf{y}_2 \\ \mathbf{y}_3 \end{bmatrix}, \quad \boldsymbol{\beta} = \begin{bmatrix} \mu_1 \\ \mu_2 \\ \mu_3 \\ \gamma \end{bmatrix} \quad \text{and} \quad \mathbf{X} = \begin{bmatrix} \mathbf{1} & \mathbf{0} & \mathbf{0} & \mathbf{x}_1 \\ \mathbf{0} & \mathbf{1} & \mathbf{0} & \mathbf{x}_2 \\ \mathbf{0} & \mathbf{0} & \mathbf{1} & \mathbf{x}_3 \end{bmatrix}$$

where **0** and **1** are vectors of length 7. Then

$$\mathbf{X}^T\mathbf{X} = \begin{bmatrix} 7 & 0 & 0 & 15 \\ 0 & 7 & 0 & 24 \\ 0 & 0 & 7 & 19 \\ 15 & 24 & 19 & 196 \end{bmatrix}, \quad \mathbf{X}^T\mathbf{y} = \begin{bmatrix} 31 \\ 53 \\ 47 \\ 398 \end{bmatrix} \quad \text{and so} \quad \mathbf{b} = \begin{bmatrix} 2.837 \\ 5.024 \\ 4.698 \\ 0.743 \end{bmatrix}.$$

Also $\mathbf{y}^T\mathbf{y} = 881$ and $\mathbf{b}^T\mathbf{X}^T\mathbf{y} = 870.698$ so for model (7.9)

$$\sigma^2 D_1 = \mathbf{y}^T\mathbf{y} - \mathbf{b}^T\mathbf{X}^T\mathbf{y} = 10.302.$$

For the reduced model (7.10)

$$\boldsymbol{\beta} = \begin{bmatrix} \mu \\ \gamma \end{bmatrix}, \quad \mathbf{X} = \begin{bmatrix} \mathbf{1} & \mathbf{x}_1 \\ \mathbf{1} & \mathbf{x}_2 \\ \mathbf{1} & \mathbf{x}_3 \end{bmatrix} \quad \text{so} \quad \mathbf{X}^T\mathbf{X} = \begin{bmatrix} 21 & 58 \\ 58 & 196 \end{bmatrix}$$

and

$$\mathbf{X}^T\mathbf{y} = \begin{bmatrix} 131 \\ 398 \end{bmatrix}.$$

Hence

$$\mathbf{b} = \begin{bmatrix} 3.447 \\ 1.011 \end{bmatrix}, \quad \mathbf{b}^T\mathbf{X}^T\mathbf{y} = 853.766$$

so $\quad \sigma^2 D_0 = 27.234.$

If we assume that model (7.9) is correct, then $D_1 \sim \chi^2_{17}$. If the null hypothesis corresponding to model (7.10) is true then $D_0 \sim \chi^2_{19}$ so

$$f = \frac{D_0 - D_1}{2} \bigg/ \frac{D_1}{17} \sim F_{2,17}.$$

For these data

$$f = \frac{16.932}{2} \bigg/ \frac{10.302}{17} = 13.97$$

indicating a significant difference in achievement scores for the training methods, after adjustment for initial differences in aptitude. The usual presentation of this analysis is given in Table 7.10.

Table 7.10 ANCOVA table for data in Table 7.9.

Source of variation	Degrees of freedom	Sum of squares	Mean square	f
Mean and covariate	2	853.766		
Factor levels	2	16.932	8.466	13.97
Residual	17	10.302	0.606	
Total	21	881.000		

7.10 Exercises

7.1 Total solids (%) were determined in each of the six batches of cream ($B_1, \ldots, B_6$) by each of three analysts (A_1, A_2 and A_3) with the results shown in the Table 7.11.

(i) Test the hypothesis H_A that there are no differences due to analysts.
(ii) Estimate the solid content of each batch. Test the hypothesis H_B that there are no differences between batches.
(iii) Examine the residuals for the most appropriate model and comment on the results of your analysis.

Table 7.11 Total solids measured in batches of cream.

	Batches					
Analysts	B_1	B_2	B_3	B_4	B_5	B_6
A_1	35.3	32.3	38.7	30.1	32.4	35.1
A_2	35.7	34.5	36.1	29.8	32.1	34.2
A_3	34.8	31.9	40.2	31.2	33.0	34.6

7.2 Perform a complete analysis of variance for the two factor experiment shown in Table 7.12. Verify that the null hypotheses of no differences due to interactions or main effects are not all independent.

Table 7.12 Example of two factor experiment.

	Factor B	
Factor A	B_1	B_2
A_1	5	3, 4
A_2	6, 4	4, 3
A_3	7	6, 8

7.3 For the achievement score data in Table 7.9:

(i) test the hypothesis that the treatment effects are equal, ignoring the covariate, i.e. compare $E(Y_{jk}) = \mu_j$ with $E(Y_{jk}) = \mu$;
(ii) test the assumption that initial aptitude has the same effect for all training methods, i.e. compare $E(Y_{jk}) = \mu_j + \gamma_j x_{jk}$ with $E(Y_{jk}) = \mu_j + \gamma x_{jk}$.

7.4 Show that $(\mathbf{y}-\mathbf{X}\boldsymbol{\beta})^{\mathrm{T}}(\mathbf{y}-\mathbf{X}\boldsymbol{\beta}) \geqslant (\mathbf{y}-\mathbf{X}\mathbf{b})^{\mathrm{T}}(\mathbf{y}-\mathbf{X}\mathbf{b})$ where $\mathbf{b}$ is any solution of the normal equations $\mathbf{X}^{\mathrm{T}}\mathbf{X}\mathbf{b} = \mathbf{X}^{\mathrm{T}}\mathbf{y}$. Hence the minimum of $(\mathbf{y}-\mathbf{X}\boldsymbol{\beta})^{\mathrm{T}}(\mathbf{y}-\mathbf{X}\boldsymbol{\beta})$ is attained when $\boldsymbol{\beta} = \mathbf{b}$ and is the same for all solutions of the normal equations.

8

BINARY VARIABLES AND LOGISTIC REGRESSION

8.1 Probability distributions

In this chapter we consider generalized linear models in which the outcome variables are measured on a binary scale. For example, the responses may be alive or dead, or present or absent. 'Success' and 'failure' are used as generic terms for the two categories.

We define the random variable

$$Y = \begin{cases} 1 & \text{if the outcome is a success} \\ 0 & \text{if the outcome is a failure} \end{cases}$$

with $\pi = \Pr(Y = 1)$ and $1 - \pi = \Pr(Y = 0)$. If there are n such random variables $Y_1, \ldots, Y_n$ which are independent with $\pi_j = \Pr(Y_j = 1)$, then their joint probability is

$$\prod_{j=1}^{n} \pi_j^{y_j}(1-\pi_j)^{1-y_j} = \exp\left[\sum_{j=1}^{n} y_j \log\left(\frac{\pi_j}{1-\pi_j}\right) + \sum_{j=1}^{n} \log(1-\pi_j)\right] \quad (8.1)$$

which is a member of the exponential family (see (3.6)).

For the case where the π_js are all equal, we define

$$R = \sum_{j=1}^{n} Y_j,$$

the number of successes in n 'trials'. The random variable R has the binomial distribution $b(n, \pi)$;

$$\Pr(R = r) = \binom{n}{r} \pi^r (1-\pi)^{n-r}, \qquad r = 0, 1, \ldots, n. \quad (8.2)$$

Therefore $E(R) = n\pi$ and var $(R) = n\pi(1-\pi)$.

In general we consider N independent random variables $R_1, R_2, \ldots, R_N$ corresponding to the numbers of successes in N different subgroups or strata (Table 8.1). If $R_i \sim b(n_i, \pi_i)$ the log-likelihood function is

$$l(\pi_1, \ldots, \pi_N; r_1, \ldots, r_N) = \sum_{i=1}^{N} \left[r_i \log\left(\frac{\pi_i}{1-\pi_i}\right) + n_i \log(1-\pi_i) + \log\binom{n_i}{r_i} \right]. \quad (8.3)$$

The distributions corresponding to (8.1) and (8.2) yield special cases of (8.3).

Table 8.1 Frequencies for N binomial distributions.

	Subgroups			
	1	2	...	N
Successes	R_1	R_2		R_N
Failures	n_1-R_1	n_2-R_2		n_N-R_N
Totals	n_1	n_2		n_N

Some of the methods discussed in this chapter use (8.3). Others use the Normal approximation to the binomial distribution,

$$R_i \sim \mathrm{N}(n_i\pi_i, n_i\pi_i(1-\pi_i)).$$

If we are mainly interested in the proportion of successes $P_i = R_i/n_i$ so that $E(P_i) = \pi_i$ and $\mathrm{var}\,(P_i) = \pi_i(1-\pi_i)/n_i$ we may use the approximate distribution $P_i \sim \mathrm{N}(\pi_i, \pi_i(1-\pi_i)/n_i)$.

8.2 Generalized linear models

The probabilities π_i are assumed to be functions of linear combinations of parameters $\boldsymbol{\beta}$, i.e.

$$g(\pi_i) = \mathbf{x}_i^{\mathrm{T}}\boldsymbol{\beta}$$

where g is the link function. The simplest case is the linear model

$$\pi_i = \mathbf{x}_i^{\mathrm{T}}\boldsymbol{\beta}.$$

This is used in some practical applications but it has the drawback that values of π_i are not restricted to the interval [0, 1]. To ensure that π has this property we often model it as a cumulative probability distribution

$$\pi = g^{-1}(\mathbf{x}^{\mathrm{T}}\boldsymbol{\beta}) = \int_{-\infty}^{t} f(z)\,\mathrm{d}z, \quad f(z) \geqslant 0,$$

where $\int_{-\infty}^{\infty} f(z)\,\mathrm{d}z = 1$. Thus $0 \leqslant \pi \leqslant 1$ and π is a non-decreasing function of t where t is related to $\mathbf{x}^{\mathrm{T}}\boldsymbol{\beta}$. The probability density function $f(z)$ is called the *tolerance distribution*. Some commonly used models are summarized in Table 8.2.

The choice of model depends on substantive and mathematical considerations. For instance, probit models have a natural interpretation for bioassay data (Finney, 1973); t represents the dose likely to kill $100\pi\%$ of experimental animals so, for example, $t = \mu$ is called the *median lethal dose* LD(50). The logistic model is particularly widely used for a variety of applications (Cox, 1970). Several of these models are illustrated in numerical examples in the next section.

Table 8.2 Some generalized linear models for binary data.

Model	Tolerance distribution $f(z)$	π	Link function
Linear	uniform on $[a, b]$	$\pi = \frac{t-a}{b-a}, \quad 0 \leqslant t \leqslant b$	$\mathbf{x}^{\mathrm{T}}\boldsymbol{\beta} = \frac{t-a}{b-a} = \pi$ (identity)
Probit	Normal $\mathrm{N}(\mu, \sigma^2)$	$\pi = \frac{1}{\sigma(2\pi)^{\frac{1}{2}}}\int_{-\infty}^{t} \exp\left\{-\frac{(z-\mu)^2}{2\sigma^2}\right\}\mathrm{d}z = \Phi\left(\frac{t-\mu}{\sigma}\right)$	$\mathbf{x}^{\mathrm{T}}\boldsymbol{\beta} = \frac{t-\mu}{\sigma} = \Phi^{-1}(\pi)$ (inverse cumulative Normal)
Logistic	$f(z) = \frac{\mathrm{e}^{(z-\mu)/k}}{k[1+\mathrm{e}^{(z-\mu)/k}]^2}$	$\pi = \frac{\mathrm{e}^{(t-\mu)/k}}{1+\mathrm{e}^{(t-\mu)/k}}$	$\mathbf{x}^{\mathrm{T}}\boldsymbol{\beta} = \frac{t-\mu}{k} = \log\left(\frac{\pi}{1-\pi}\right)$ (logit)
Extreme value	$f(z) = \frac{1}{b}\exp\left[\frac{z-a}{b} - \mathrm{e}^{(z-a)/b}\right]$	$\pi = 1-\exp\left[-\exp\left(\frac{t-a}{b}\right)\right]$	$\mathbf{x}^{\mathrm{T}}\boldsymbol{\beta} = \frac{t-a}{b} = \log(-\log(1-\pi))$ (complementary log log)

8.3 Maximum likelihood estimation and the log-likelihood ratio statistic

Maximum likelihood estimates of the parameters $\boldsymbol{\beta}$ and consequently of the probabilities $\pi_i = g(\mathbf{x}_i^{\mathrm{T}}\boldsymbol{\beta})$, $i = 1, \ldots, N$ are obtained by maximizing the log-likelihood function (8.3) using the methods described in Chapter 4. To measure the goodness of fit of a model we use

$$D = 2[l(\hat{\boldsymbol{\pi}}_{\max}; \mathbf{r}) - l(\hat{\boldsymbol{\pi}}_0; \mathbf{r})]$$

where $\mathbf{r} = [r_1, \ldots, r_N]^{\mathrm{T}}$, $\hat{\boldsymbol{\pi}}_0 = [\hat{\pi}_1, \ldots, \hat{\pi}_N]^{\mathrm{T}}$ is the vector of maximum likelihood estimates corresponding to the model and $\hat{\boldsymbol{\pi}}_{\max}$ is the vector of maximum likelihood estimates for the maximal model.

Without loss of generality, for the maximal model we take the π_is as the parameters to be estimated. Then

$$\frac{\partial l}{\partial \pi_i} = \frac{r_i}{\pi_i} - \frac{n_i - r_i}{1 - \pi_i}$$

so the ith element of $\hat{\boldsymbol{\pi}}_{\max}$, the solution of $\partial l / \partial \pi_i = 0$, is r_i / n_i the observed proportion of successes in the ith subgroup. Hence

$$l(\hat{\boldsymbol{\pi}}_{\max}; \mathbf{r}) = \sum_{i=1}^{N} \left[r_i \log\left(\frac{r_i}{n_i}\right) + (n_i - r_i) \log\left(\frac{n_i - r_i}{n_i}\right) + \log\binom{n_i}{r_i} \right]$$

and so

$$D = \sum_{l=1}^{N} \left[r_i \log\left(\frac{r_i}{n\hat{\pi}_i}\right) + (n_i - r_i) \log\left(\frac{n_i - r_i}{n_i - n_i \hat{\pi}_i}\right) \right]. \tag{8.4}$$

Thus D has the form

$$D = 2\Sigma\, o \log \frac{o}{e}$$

where o denotes the observed frequencies r_i and $n_i - r_i$ in the cells of Table 8.1, e denotes the corresponding estimated expected frequencies (since $E(R_i) = n_i \pi_i$) and summation is over all $2 \times N$ cells.

For a model involving $p < N$ independent parameters the fit is assessed by comparing D in (8.4) with critical values of the χ^2_{N-p} distribution. This is illustrated in the following example.

Example 8.1 *Dose–response relationships*

In bioassays the response may vary with a covariate termed the *dose*. A typical example involving a binary response is given in Table 8.3, where R is the number of beetles killed after 5 h exposure to gaseous carbon disulphide at various concentrations (data from Bliss, 1935). Figure 8.1 shows the proportions $p_i = r_i / n_i$ plotted against dosage x_i.

Table 8.3 Beetle mortality data

Dose x_i ($\log_{10} CS_2$ mg l^{-1})	Number of insects, n_i	Number killed, r_i
1.6907	59	6
1.7242	60	13
1.7552	62	18
1.7842	56	28
1.8113	63	52
1.8369	59	53
1.8610	62	61
1.8839	60	60

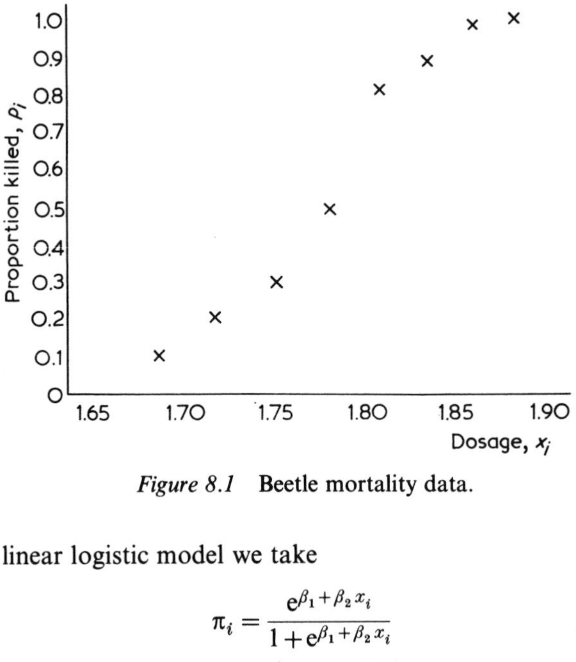

Figure 8.1 Beetle mortality data.

For the linear logistic model we take

$$\pi_i = \frac{e^{\beta_1+\beta_2 x_i}}{1+e^{\beta_1+\beta_2 x_i}}$$

so the link function is the *logit* defined as the logarithm of the odds $(\pi_i/1-\pi_i)$

$$\text{logit}\,(\pi_i) = \log\left(\frac{\pi_i}{1-\pi_i}\right) = \beta_1+\beta_2 x_i.$$

From (8.3) the log-likelihood function is

$$l = \sum_{i=1}^{N}\left[r_i(\beta_1+\beta_2 x_i) - n_i \log(1+e^{\beta_1+\beta_2 x_i}) + \log\binom{n_i}{r_i}\right]$$

and the scores with respect to β_1 and β_2 are:

$$U_1 = \frac{\partial l}{\partial \beta_1} = \Sigma\left[r_i - n_i\left(\frac{e^{\beta_1+\beta_2 x_i}}{1+e^{\beta_1+\beta_2 x_i}}\right)\right] = \Sigma\,(r_i - n_i\,\pi_i),$$

$$U_2 = \frac{\partial l}{\partial \beta_2} = \Sigma\left[r_i x_i - n_i x_i\left(\frac{e^{\beta_1+\beta_2 x_i}}{1+e^{\beta_1+\beta_2 x_i}}\right)\right] = \Sigma\, x_i(r_i - n_i\,\pi_i).$$

Similarly the information matrix is

$$\mathscr{I} = \begin{bmatrix} \Sigma\, n_i\,\pi_i(1-\pi_i) & \Sigma\, n_i\, x_i\,\pi_i(1-\pi_i) \\ \Sigma\, n_i\, x_i\,\pi_i(1-\pi_i) & \Sigma\, n_i\, x_i^2\,\pi_i(1-\pi_i) \end{bmatrix}.$$

Maximum likelihood estimates are obtained by solving the iterative equation

$$\mathscr{I}^{(m-1)}\mathbf{b}^{(m)} = \mathscr{I}^{(m-1)}\mathbf{b}^{(m-1)} + \mathbf{U}^{(m-1)}$$

(from (4.7)) where m indicates the mth approximation and $\mathbf{b} = [b_1\, b_2]^{\mathrm{T}}$ is the vector of estimates. Starting from $b_1^{(0)} = b_2^{(0)} = 0$ successive approximations are shown in Table 8.4 together with the estimated frequencies $\hat{r}_i = n_i\,\hat{\pi}_i$, the estimated variance–covariance matrix $[\mathscr{I}(\mathbf{b})]^{-1}$ and the log-likelihood ratio statistic.

Table 8.4 Fitting the linear logistic model to the beetle mortality data.

		Initial estimate	First approx.	Second approx.	Fourth approx.	Tenth approx.
	b_1	0	−37.849	−53.851	−60.700	−60.717
	b_2	0	21.334	30.382	34.261	34.270
Observations		*Fitted values*				
r_1	6	29.5	8.508	4.544	3.460	3.458
r_2	13	30.0	15.369	11.254	9.845	9.842
r_3	18	31.0	24.810	23.059	22.454	22.451
r_4	28	28.0	30.983	32.946	33.896	33.898
r_5	52	31.5	43.361	48.197	50.092	50.096
r_6	53	29.5	46.739	51.704	53.288	53.291
r_7	61	31.0	53.593	58.060	59.220	59.222
r_8	60	30.0	54.732	58.036	58.742	58.743

$$[\mathscr{I}(\mathbf{b})]^{-1} = \begin{bmatrix} 26.802 & 15.061 \\ 15.061 & 8.469 \end{bmatrix}, \qquad D = 11.23$$

The standard errors of the estimates $b_1 = -60.72$ and $b_2 = 34.27$ are $(26.802)^{\frac{1}{2}} = 5.18$ and $(8.469)^{\frac{1}{2}} = 2.91$, respectively. Under the null hypothesis that the linear logistic model describes the data, D has an approximate χ_6^2 distribution because there are $N = 8$ dosage groups and $p = 2$ parameters.

But the upper 5% point of the χ^2_6 distribution is 12.59 which indicates that the model does not fit the data particularly well.

Using the computer program GLIM (Baker and Nelder, 1978) the probit model $\pi = \Phi(\beta_1 + \beta_2 x_i)$ and the extreme value model $\pi = 1 - \exp(-\exp(\beta_1 + \beta_2 x_i))$ were also fitted to these data. The results are shown in Table 8.5. Among these models the extreme value model clearly provides the best description of the data.

Table 8.5 Comparison of various dose–response models for the beetle mortality data.

Observed value of R	Logistic model	Probit model	Extreme value model
6	3.46	3.36	5.59
13	9.84	10.72	11.28
18	22.45	23.48	20.95
28	33.90	33.82	30.37
52	50.10	49.62	47.78
53	53.29	53.32	54.14
61	59.22	59.66	61.11
60	58.74	59.23	59.95
D	11.23	10.12	3.45

8.4 General logistic regression

The simple linear logistic model $\log[\pi_i/(1-\pi_i)] = \beta_1 + \beta_2 x_i$ used in the above example is a special case of the general logistic regression model

$$\text{logit}\,\pi_i = \log\left(\frac{\pi_i}{1-\pi_i}\right) = \mathbf{x}_i^{\mathrm{T}}\boldsymbol{\beta}$$

where $\mathbf{x}_i$ is a vector of continuous measurements corresponding to covariates and dummy variables corresponding to factor levels and $\boldsymbol{\beta}$ is the parameter vector.

This model is very widely used for analysing multivariate data involving binary responses because it is a powerful technique analogous to multiple regression and ANOVA for continuous responses. Computer programs for performing logistic regression are available in most statistical packages (e.g. the program PLR in BMDP).

There are several points that should be made about the use of logistic regression. First, maximum likelihood estimation can be performed even when there is only a single response, i.e. $n_i = 1$ and $R_i = 0$ or 1, for each parameter combination $\mathbf{x}_i^{\mathrm{T}}\boldsymbol{\beta}$. Second, for logistic regression the method of maximum likelihood involves iteration and so various alternative procedures, based on the method of least squares, are often used to reduce the amount of computation required. Third, as illustrated in the dose–response example (Example 8.1), the logistic model may not fit the data as well as some alternative model so it is important to assess the overall goodness of fit of the model as well as the relative contributions due to various parameters. (Notice that for the binomial distribution the log-likelihood ratio statistic D is completely determined, unlike for the Normal distribution where it depends on the unknown parameter σ^2.)

The general use of logistic regression for more complicated data is illustrated in Example 8.2 which also involves comparisons with other models.

8.5 Other criteria for goodness of fit

Instead of using maximum likelihood estimation one could estimate the parameters by minimizing the weighted sum of squares

$$S_w = \sum_{i=1}^{N} \frac{(r_i - n_i \pi_i)^2}{n_i \pi_i (1-\pi_i)}$$

since $E(R_i) = n_i \pi_i$ and $\mathrm{var}(R_i) = n_i \pi_i (1-\pi_i)$.

This is equivalent to minimizing the *Pearson chi-squared statistic*

$$X^2 = \sum \frac{(o-e)^2}{e}$$

where o represents the observed frequencies in Table 8.1, e represents the expected frequencies under the model and summation is over all $2 \times N$ cells of the table. The reason is

$$\begin{aligned} X^2 &= \sum_{i=1}^{N} \frac{(r_i - n_i \pi_i)^2}{n_i \pi_i} + \sum_{i=1}^{N} \frac{[n_i - r_i - n_i(1-\pi_i)]^2}{n_i(1-\pi_i)} \\ &= \sum_{i=1}^{N} \frac{(r_i - n_i \pi_i)^2}{n_i \pi_i (1-\pi_i)} (1 - \pi_i + \pi_i) = S_w. \end{aligned}$$

When X^2 is evaluated at the estimated expected frequencies, the statistic

$$X^2 = \sum_{i=1}^{N} \frac{(r_i - n_i \hat{\pi}_i)^2}{n_i \hat{\pi}_i (1-\hat{\pi}_i)}$$

is asymptotically equivalent to the log-likelihood ratio statistic in (8.4).

$$D = 2 \sum_{i=1}^{N} \left[r_i \log\left(\frac{r_i}{n_i \hat{\pi}_i}\right) + (n_i - r_i) \log\left(\frac{n_i - r_i}{n_i - n_i \hat{\pi}_i}\right) \right].$$

The proof uses the Taylor series expansion of $x \log(x/y)$ about $x = y$, namely,

$$x \log \frac{x}{y} = (x-y) + \tfrac{1}{2} \frac{(x-y)^2}{y} + \ldots.$$

Thus

$$D = 2\sum_{i=1}^{N}\left[(r_i - n_i\hat{\pi}_i) + \tfrac{1}{2}\frac{(r_i - n_i\hat{\pi}_i)^2}{n_i\hat{\pi}_i}\right.$$

$$\left. + (n_i - r_i - n_i + n_i\hat{\pi}_i) + \tfrac{1}{2}\frac{(n_i - r_i - n_i + n_i\hat{\pi}_i)^2}{n_i - n_i\hat{\pi}_i} + \ldots\right]$$

$$\simeq \sum_{i=1}^{N}\frac{(r_i - n_i\hat{\pi}_i)^2}{n_i\hat{\pi}_i(1-\hat{\pi}_i)} = X^2.$$

The large sample distribution of D, under the null hypothesis that the model is correct, is $D \sim \chi^2_{N-p}$ therefore approximately

$$X^2 \sim \chi^2_{N-p}.$$

Another criterion for goodness of fit is the *modified chi-squared statistic* obtained by replacing the estimated probabilities in the denominator of X^2 by the relative frequencies,

$$X^2_{\text{mod}} = \sum_{i=1}^{N}\frac{(r_i - n_i\hat{\pi}_i)^2}{r_i(n_i - r_i)/n_i}.$$

Asymptotically this too has the χ^2_{N-p} distribution if the model is correct.

The choice between D, X^2 and X^2_{mod} depends on the adequacy of the approximation to the χ^2_{N-p} distribution. There is some evidence to suggest that X^2 is better than D (Larntz, 1978) because D is unduly influenced by very small frequencies. All the approximations are likely to be poor if the expected frequencies are too small (e.g. less than 1).

8.6 Least squares methods

There are some computational advantages in using weighted least squares estimation instead of maximum likelihood.

Consider a function $\psi(P_i)$ of the proportion of successes in the ith subgroup. The Taylor series expansion of $\psi(P_i)$ about $P_i = \pi_i$ is

$$\psi(P_i) = \psi\left(\frac{R_i}{n_i}\right) = \psi(\pi_i) + \left(\frac{R_i}{n_i} - \pi_i\right)\psi'(\pi_i) + o\left(\frac{1}{n_i^2}\right).$$

Thus, to a first approximation,

$$E[\psi(P_i)] = \psi(\pi_i)$$

since $E(R_i/n_i) = \pi_i$. Also

$$\begin{aligned}\text{var}[\psi(P_i)] &= E[\psi(P_i) - \psi(\pi_i)]^2 \\ &= [\psi'(\pi_i)]^2 E\left[\frac{R_i}{n_i} - \pi_i\right]^2 \\ &= [\psi'(\pi_i)]^2 \frac{\pi_i(1-\pi_i)}{n_i}\end{aligned}$$

since

$$E\left(\frac{R_i}{n_i}-\pi_i\right)^2 = \text{var}(P_i) = \pi_i(1-\pi_i)/n_i.$$

Hence the weighted least squares criterion is

$$X^2 = \sum_{i=1}^{N} \frac{[\psi(r_i/n_i)-\psi(\pi_i)]^2}{[\psi'(\pi_i)]^2\pi_i(1-\pi_i)/n_i}.$$

Some common choices of ψ are summarized in Table 8.6 and discussed below.

Table 8.6 Some weighted least squares models for binary data.

$\psi(\pi_i)$	X^2
π_i	$\Sigma\dfrac{(p_i-\pi_i)^2}{\pi_i(1-\pi_i)/n_i}$
logit π_i	$\Sigma[(\text{logit}\,p_i-\text{logit}\,\pi_i)^2\,\pi_i(1-\pi_i)n_i]$
$\sin^{-1}(\pi_i)^{\frac{1}{2}}$	$\Sigma\,4n_i(\sin^{-1}(p_i)^{\frac{1}{2}}-\sin^{-1}(\pi_i)^{\frac{1}{2}})^2$

First, if $\psi(\pi_i) = \pi_i$ and $\pi_i = \mathbf{x}_i^T\boldsymbol{\beta}$ the modified X^2 criterion is

$$X^2_{\text{mod}} = \sum_{i=1}^{N}\frac{(p_i-\mathbf{x}_i^T\boldsymbol{\beta})^2}{p_i(1-p_i)/n_i}, \tag{8.5}$$

which is linear in $\boldsymbol{\beta}$ so estimation does not involve any iteration. However, the estimates $\hat{\pi}_i = \mathbf{x}_i^T\mathbf{b}$ may not lie between 0 and 1.

Second, if $\psi(\pi_i) = \text{logit}\,\pi_i$ and $\pi_i = e^{\mathbf{x}_i^T\boldsymbol{\beta}}/(1+e^{\mathbf{x}_i^T\boldsymbol{\beta}})$ then

$$X^2_{\text{mod}} = \sum_{i=1}^{N}(z_i-\mathbf{x}_i^T\boldsymbol{\beta})^2\frac{r_i(n_i-r_i)}{n_i}, \tag{8.6}$$

where

$$z_i = \text{logit}\,p_i = \log\left(\frac{r_i}{n_i-r_i}\right).$$

This involves no iteration and yields estimates of the π_is in the range [0, 1]. Cox (1970) calls this the *empirical logistic transformation* and recommends the use of

$$Z_i = \log\left(\frac{R_i+\frac{1}{2}}{n_i-R_i+\frac{1}{2}}\right)$$

instead of

$$Z_i = \log\left(\frac{R_i}{n_i-R_i}\right)$$

to reduce the bias $E(Z_i-\mathbf{x}_i^T\boldsymbol{\beta})$ (see Exercise 8.3). The minimum value of (8.6) is called the *minimum logit chi-squared statistic* (Berkson, 1953).

Third, the *arc sin transformation*, $\psi(\pi_i) = \sin^{-1}(\pi_i)^{\frac{1}{2}}$ (with any choice of π_i), is said to have the *variance stabilizing* property because

$$\mathrm{var}\,[\psi(P_i)] = [\psi'(\pi_i)]^2\pi_i(1-\pi_i)/n_i = (4n_i)^{-1}.$$

Thus the weight does not depend on the parameters or the responses.

In the following example we use logistic regression models (with maximum likelihood estimation) and linear models (with weighted least squares estimation) to analyse some binary data involving a covariate and a qualitative factor.

Table 8.7 Anther data.

		Centrifuging force (g)		
		40	150	350
Control	p_{1k}	0.539	0.525	0.528
	n_{1k}	102	99	108
Treatment	p_{2k}	0.724	0.617	0.555
	n_{2k}	76	81	90

Example 8.2

These data (Table 8.7), cited by Wood (1978) are taken from Sangwan-Norrell (1977). They are proportions p_{jk} of embryogenic anthers of the plant species *Datura innoxia* Mill. obtained when numbers n_{jk} of anthers were prepared under several different conditions. There is one qualitative factor, a treatment consisting of storage at 3 °C for 48 h or a control storage condition, and a covariate, three values of centrifuging force. We will compare the treatment and control effects on the proportions after adjustment (if necessary) for centrifuging force.

The proportions in the control and treatment groups are plotted against x_k, the logarithm of the centrifuging force, in Fig. 8.2. The response proportions appear to be higher in the treatment group than in the control group and, at least for the treated group, the response decreases with x_k. We use the logistic and linear models shown in Table 8.8.

For models 1(a) and 1(b) the slopes and intercepts differ between the treatment and control groups, and for the linear components $\mathbf{x}^T\boldsymbol{\beta}$, $\mathbf{x}_{1k}^T = [1, 0, x_k, 0]$, $\mathbf{x}_{2k}^T = [0, 1, 0, x_k]$ and $\boldsymbol{\beta} = [\alpha_1, \alpha_2, \beta_1, \beta_2]^T$. For models 2(a) and 2(b) the intercepts differ but not the slope so $\mathbf{x}_{1k}^T = [1, 0, x_k]$, $\mathbf{x}_{2k}^T = [0, 1, x_k]$ and $\boldsymbol{\beta} = [\alpha_1, \alpha_2, \beta]^T$. For models 3(a) and 3(b) the intercept and slope are the same for both groups so $\mathbf{x}_k^T = [1, x_k]$ and $\boldsymbol{\beta} = [\alpha, \beta]^T$.

The logistic models were fitted by the method of maximum likelihood using GLIM. The results are summarized in Table 8.9.

Table 8.8 Models for anther data ($j = 1, 2$; $k = 1, 2, 3$).

	logit $\pi = \mathbf{x}^T\boldsymbol{\beta}$		$\pi = \mathbf{x}^T\boldsymbol{\beta}$
1(a)	$\text{logit}\,\pi_{jk} = \alpha_j + \beta_j x_k$	1(b)	$\pi_{jk} = \alpha_j + \beta_j x_k$
2(a)	$\text{logit}\,\pi_{jk} = \alpha_j + \beta x_k$	2(b)	$\pi_{jk} = \alpha_j + \beta x_k$
3(a)	$\text{logit}\,\pi_{jk} = \alpha + \beta x_k$	3(b)	$\pi_{jk} = \alpha + \beta x_k$

Table 8.9 Maximum likelihood estimation for logistic models for anther data (standard errors of estimates in brackets).

1(a)	2(a)	3(a)
$a_1 = 0.238$ (0.628)	$a_1 = 0.877$ (0.487)	$a = 1.021$ (0.481)
$a_2 - a_1 = 1.977$ (0.998)	$a_2 - a_1 = 0.407$ (0.175)	$b = -0.148$ (0.096)
$b_1 = -0.023$ (0.127)	$b = -0.155$ (0.097)	
$b_2 - b_1 = -0.319$ (0.199)		
$D_1 = 0.0277$	$D_2 = 2.619$	$D_3 = 8.092$

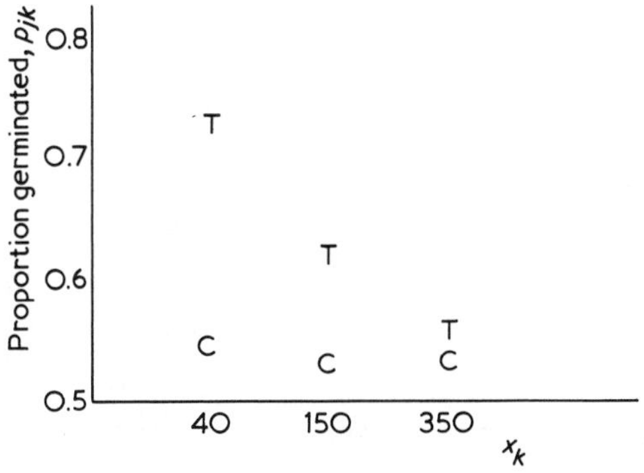

Figure 8.2 Plot of anther data in Table 8.7. C and T indicate control and treatment conditions, respectively.

For model 1(a) D_1 is very small (compared with the χ^2_2 distribution) indicating a good fit which is not surprising since the model has $p = 4$ parameters to describe $N = 6$ data points. However, the more parsimonious models 2(a) and 3(a) do not fit as well (the upper 5% points of χ^2_2, χ^2_3 and χ^2_4 corresponding to the three models are 5.99, 7.81 and 9.49, respectively).

To test the null hypothesis that the slope is the same for both the treatment and control groups we use $D_2 - D_1 = 2.591$. Since $0.1 < \Pr(\chi^2_1 < 2.591) < 0.2$ we could conclude that the data provide little evidence against the null hypothesis of equal slopes. On the other hand, the power of this test is very low and both Fig. 8.2 and the estimates for model 1(a) suggest that although the slope for the control group may be zero, the slope for the treatment group is negative. Comparison of the log-likelihood statistics from models 2(a) and

3(a) gives a test for equality of the control and treatment effects after a common adjustment for centrifuging force: $D_3 - D_2 = 5.473$, which is very significant compared with χ_1^2 so we would conclude that the storage effects differ.

The linear models were fitted by weighted least squares using the expression in (8.5). The results are summarized in Table 8.10.

Table 8.10 Weighted least squares estimation for anther data.

1(b)	2(b)	3(b)
$a_1 = 0.558$	$a_1 = 0.729$	$a = 0.776$
$a_2 = 1.010$	$a_2 = 0.831$	$b = -0.041$
$b_1 = -0.006$	$b = -0.041$	
$b_2 = -0.078$		
$X^2 = 0.017$	$X^2 = 2.362$	$X^2 = 8.449$

Table 8.11 Observed proportions and probabilities estimated from logistic and linear models for anther data.

	Logistic models		Linear models	
Observed proportions	1(a)	2(a)	1(b)	2(b)
0.539	0.537	0.576	0.537	0.579
0.525	0.530	0.526	0.530	0.525
0.528	0.525	0.493	0.525	0.490
0.724	0.721	0.671	0.723	0.680
0.617	0.623	0.625	0.620	0.627
0.555	0.553	0.593	0.554	0.592
	$D = 0.028$	$D = 2.619$	$X^2 = 0.017$	$X^2 = 2.362$

For these data the logistic models (fitted by the maximum likelihood method) and the analogous linear models (fitted by the method of weighted least squares) give remarkably similar results – see Table 8.11. The reason is that all the observed proportions are near $\frac{1}{2}$ and in the neighbourhood of $y = e^z/(1+e^z) = \frac{1}{2}$ (i.e. near $z = 1$) the logistic function is approximately linear. If the proportions had been near 0 or 1 the models would have differed much more.

The interpretation of these results (for either method of analysis) is that the two storage methods lead to significantly different proportions of embryogenic anthers and that, at least for the treatment storage condition, the proportions decrease with increasing centrifuging force.

8.7 Remarks

Many of the issues that arise in the use of multiple regression for continuous response variables are also relevant with binary responses. Tests for the inclusion or exclusion of certain terms usually are not independent and it is necessary to state carefully which terms are included in the model at each stage. If there are many explanatory variables, stepwise selection methods can be used to identify best subsets of variables.

Graphical examination of residuals is useful for assessing the adequacy of a proposed model. A simple definition for standardized residuals is

$$d_i = \frac{p_i - \hat{\pi}_i}{(\hat{\pi}_i(1-\hat{\pi}_i)/n_i)^{\frac{1}{2}}}$$

where $p_i = r_i/n_i$ is the observed proportion and $\hat{\pi}_i$ is the proportion estimated under the model. The d_is approximately have zero mean and unit variance and when plotted against factor levels and covariates should not show any patterns. However, their probability distribution may be far from Normal. More complicated residuals, which are nearly Normal, are defined by Cox and Snell (1968).

8.8 Exercises

8.1 The table overleaf gives data reported by Gordon and Foss (1966). On each of 18 days very young babies in a hospital nursery were chosen as subjects if they were not crying at a certain instant. One baby selected at random was rocked for a set period, the remainder serving as controls. The numbers not crying at the end of a specified period are given in Table 8.12. (There is no information about the extent to which the same infant enters the experiment on a number of days so we will treat responses on different days as independent.)

(i) Pool the data from the different days into a single 2×2 contingency table and test the hypothesis that the probability of crying is the same for rocked and unrocked babies.

(ii) The analysis in (i) ignores the matching by days. To incorporate this aspect, re-analyse the data using a logistic model with parameters for days and control or experimental conditions. How well does it fit the data? Examine the residuals to see if there are any patterns in the data which are not accounted for by the model. By fitting a model which ignores the control or experimental effects, test the hypothesis that rocking does not affect the probability of crying. What is the simplest model which describes the data well? (Cox, 1966, 1970)

Table 8.12 Crying babies data.

Day	No. of control babies	No. not crying	No. of experimental babies	No. not crying
1	8	3	1	1
2	6	2	1	1
3	5	1	1	1
4	6	1	1	0
5	5	4	1	1
6	9	4	1	1
7	8	5	1	1
8	8	4	1	1
9	5	3	1	1
10	9	8	1	0
11	6	5	1	1
12	9	8	1	1
13	8	5	1	1
14	5	4	1	1
15	6	4	1	1
16	8	7	1	1
17	6	4	1	0
18	8	5	1	1

8.2 *Odds ratios.* Consider a 2×2 contingency table from a prospective study in which people who were or were not exposed to some pollutant are followed-up and, after several years, categorized according to the presence or absence of a disease. Table 8.13 shows the probabilities for each cell.

Table 8.13 Probabilities for a prospective study.

	Diseased	Not diseased
Exposed	π_1	$1-\pi_1$
Not exposed	π_2	$1-\pi_2$

The odds of disease for either exposure group is $O_i = \dfrac{\pi_i}{1-\pi_i} (i = 1, 2)$ and so the odds ratio

$$\psi = \frac{O_1}{O_2} = \frac{\pi_1(1-\pi_2)}{\pi_2(1-\pi_1)}$$

is a measure of the relative likelihood of disease for the exposed and not exposed groups.

(i) For the simple logistic model $\pi_i = e^{\alpha_i}/(1+e^{\alpha_i})$ show that $\psi = 1$ corresponds to no difference on the logit scale between the exposed and not exposed groups.

(ii) Consider N 2×2 tables like Table 8.13, one for each level x_j $(j = 1, \ldots, N)$ of a factor (e.g. age groups). For the logistic model

$$\pi_{ij} = \frac{e^{\alpha_i+\beta_i x_j}}{1+e^{\alpha_i+\beta_i x_j}}, \quad i = 1,2; \quad j = 1,\ldots,N.$$

Show that $\log \psi$ is constant over all tables if $\beta_1 = \beta_2$ (i.e., the πs are parallel on the logit scale). (McKinlay, 1978)

8.3 Let the random variable R have the binomial distribution with parameters n and π and consider the transformation $\psi[(R+a)/(n+b)]$ where a and b are constants.

(i) Use the Taylor expansion of $\psi[(R+a)/(n+b)]$ about $[(R+a)/(n+b)] = \pi$ and the approximation

$$\frac{R+a}{n+b} - \pi = \frac{1}{n}[(R-n\pi)+(a-b\pi)]\left[1-\frac{b}{n}+\left(\frac{b}{n}\right)^2-\ldots\right]$$

to show that

$$E\left[\psi\left(\frac{R+a}{n+b}\right)\right] = \psi(\pi) + \frac{\psi'(\pi)(a-b\pi)}{n} + \frac{\psi''(\pi)\pi(1-\pi)}{2n} + o\left(\frac{1}{n^2}\right)$$

and

$$\operatorname{var}\left[\psi\left(\frac{R+a}{n+b}\right)\right] = [\psi'(\pi)]^2\frac{\pi(1-\pi)}{n} + o\left(\frac{1}{n^2}\right).$$

(ii) For $\psi(t) = \log[t/(1-t)]$ show that

$$\text{bias} = E\left[\psi\left(\frac{R+a}{n+b}\right) - \psi(\pi)\right]$$

is of order n^{-2} if $a = \frac{1}{2}$ and $b = 1$, i.e., the empirical logistic transform $\log[(R+\frac{1}{2})/(n-R+\frac{1}{2})]$ is less biased than $\log[R/n(-R)]$.

(iii) For the log transform $\psi(t) = \log t$ find a and b to reduce the bias to $o(n^{-2})$ and find the variance. (Cox, 1970)

8.4 *Dose–response models*: Wider classes of models than those described in Section 8.3 have been proposed for dose–response relationships.

(i) Let

$$\pi = \frac{1}{\beta(m_1, m_2)}\int_{-\infty}^{t} e^{m_1 z}(1+e^z)^{-(m_1+m_2)}dz$$

where $\beta(m_1, m_2) = \Gamma(m_1)\Gamma(m_2)/\Gamma(m_1+m_2)$ is the beta function. Show that the logistic model is given by $m_1 = m_2 = 1$. Show that as $m_1 \to \infty$ and $m_2 \to \infty$, π converges to the probit model. (Prentice, 1976)

(ii) Let

$$t = \int_{\frac{1}{2}}^{\pi} \frac{dz}{1 - |2z-1|^{\nu+1}}, \quad \nu > -1.$$

Show that $\nu = 1$ gives the linear logistic model, and as $\nu \to \infty$, π converges to the uniform model. (This generalization allows for symmetric tolerance distributions with heavier or lighter tails than those discussed in Section 8.3 and may provide better estimates for extreme percentiles.)

(Copenhaver and Mielke, 1977)

(iii) Let

$$\pi = \begin{cases} 1-(1+\lambda e^t)^{-1/\lambda}, & \lambda e^t > -1 \\ 1, & \text{otherwise.} \end{cases}$$

Show that the logistic model is given by $\lambda = 1$ and that $\lambda = 0$ corresponds to the extreme value model.

(Aranda-Ordaz, 1981)

9

CONTINGENCY TABLES AND LOG-LINEAR MODELS

9.1 Introduction

This chapter is about the analysis of data in which the response and explanatory variables are all categorical, i.e. they are measured on nominal or possibly ordinal scales. Each scale may have more than two categories. Unlike the methods described in previous chapters, generalized linear models for categorical data can readily be defined when there is more than one response variable. The observations consist of *counts* or *frequencies* in the cells of a *contingency table* formed by the cross-classification of response and explanatory variables.

We begin with three numerical examples representing different study designs. For each we consider the roles of the various variables and identify the relevant hypotheses.

Example 9.1 *Tumour type and site*

These data are from a *cross-sectional study* of patients with a form of skin cancer called malignant melanoma. For a sample of $n = 400$ patients the site of the tumour and its histological type were recorded. The data, numbers of patients with each combination of tumour type and site, are given in Table 9.1.

In this example there are two response variables, site and histological type. The cell frequencies are regarded as random variables which are subject to the constraint that they must add to n.

The question of interest is whether there is any association between the two response variables. Table 9.2 shows the data displayed as percentages of row and column totals. It appears that Hutchinson's melanotic freckle is more common on the head and neck but there is little evidence of associations between other tumour types and sites.

Table 9.1 Tumour type and site: frequencies (Roberts *et al.*, 1981).

Histological type	Site: Head and neck	Trunk	Extremities	Total
Hutchinson's melanotic freckle	22	2	10	34
Superficial spreading melanoma	16	54	115	185
Nodular	19	33	73	125
Indeterminate	11	17	28	56
Total	68	106	226	400

Table 9.2 Tumour type and site: percentages of row or column totals.

Histological types	Site: Head and neck	Trunk	Extremities	All sites
		Row percentages		
Hutchinson's melanotic freckle	64.7	5.9	29.4	100
Superficial spreading melanoma	8.6	29.2	62.2	100
Nodular	15.2	26.4	58.4	100
Indeterminate	19.6	30.4	50.0	100
All types	17.0	26.5	56.5	100
		Column percentages		
Hutchinson's melanotic freckle	32.4	1.9	4.4	8.50
Superficial spreading melanoma	23.5	50.9	50.9	46.25
Nodular	27.9	31.1	32.3	31.25
Indeterminate	16.2	16.0	12.4	14.00
All types	100.0	99.9	100.0	100.0

Example 9.2 *Randomized controlled trial*

In a prospective study of a new drug, patients were randomly allocated to three groups each of 50 people. Two groups were given the new drug at different dosage levels and the third group received a placebo. The responses of all patients are given in Table 9.3 (these are artificial data).

For this study there is one explanatory variable, treatment, and one response variable. The cell frequencies in each row of the table are constrained to add to 50. We want to know if the pattern of responses is the same at each level of the treatment.

Table 9.3 Randomized controlled trial.

Treatment	Response: Improved	No difference	Worse	Total
Placebo	18	17	15	50
Half dose	20	10	20	50
Full dose	25	13	12	50

Example 9.3 *Peptic ulcers and blood groups*

In this retrospective *case–control study* a group of peptic ulcer patients was assembled and a group of control patients not known to have peptic ulcer who presumably were similar to ulcer patients with respect to age, sex and perhaps some other variables. Blood groups were ascertained for all subjects and the study was replicated in three cities. The results are shown in Table 9.4.

Table 9.4 Peptic ulcers and blood groups: frequencies (Woolf, 1955).

		Blood groups: A	O	Total
London	Cases	579	911	1490
	Controls	4219	4578	8797
Manchester	Cases	246	361	607
	Controls	3775	4532	8307
Newcastle	Cases	291	396	687
Upon Tyne	Controls	5261	6598	11859

This is a $2 \times 2 \times 3$ contingency table with one response variable (blood group) and two explanatory variables. In the subtable for each city the row totals for cases and controls are presumed to be fixed. The relevant questions include:

(i) Is the distribution of blood groups the same in all cities?
(ii) In each city is there an association between blood group and peptic ulcer?
(iii) Is any such association the same in all cities?

When the data are presented as percentages of row totals (Table 9.5) it appears that blood group A is less common among ulcer patients than among controls and that the blood group distributions may differ between the cities.

Table 9.5 Peptic ulcers and blood groups: percentages.

		Blood groups		
		A	O	Total
London	Cases	38.9	61.1	100
	Controls	48.0	52.0	100
Manchester	Cases	40.5	59.5	100
	Controls	45.4	54.6	100
Newcastle	Cases	42.4	57.6	100
Upon Tyne	Controls	44.4	55.6	100

This chapter concerns generalized linear models for categorical data when the contingency tables have relatively simple structure. We ignore complicated situations, for example, when some cells of the table necessarily have zero frequencies (e.g. male hysterectomy cases) or when the responses can be regarded as repeated measures on the same individuals. For more complete treatment of contingency tables the reader is referred to the books by Bishop *et al.* (1975), Everitt (1977) or Fienberg (1977).

9.2 Probability distributions

For two-dimensional tables with J categories for variable A and K categories for variable B we use the notation in Table 9.6 in which Y_{jk} denotes the frequency for the (j, k)th cell, $Y_{j.}$ and $Y_{.k}$ denote the row and column totals and n the overall total. The cell frequencies Y_{jk} are the dependent variables we wish to model.

In general for a $J \times K \times \ldots \times L$ table we write the frequencies $Y_{jk\ldots l}$ in a single vector $\mathbf{y}$ with elements indexed by $i = 1, \ldots, N$.

We begin with probability models for two-dimensional tables. The simplest is obtained by assuming that the random variables Y_{jk} are independent and each has the *Poisson distribution* with parameter $\lambda_{jk} \geqslant 0$. Their joint distribution is

$$f(\mathbf{y}; \boldsymbol{\lambda}) = \prod_{j=1}^{J} \prod_{k=1}^{K} \frac{\lambda_{jk}^{y_{jk}} e^{-\lambda_{jk}}}{y_{jk}!}.$$

More commonly there are constraints on the Y_{jk}s, for example, that the total frequency n is fixed by the study design. In this case, from the additive property of independent Poisson variates, n has the Poisson distribution with parameter $\lambda = \Sigma\Sigma\lambda_{jk}$. Therefore the joint distribution of the Y_{jk}s, conditional on n, is

$$f(\mathbf{y}|n) = \frac{n!}{e^{-\lambda}\lambda^n} \prod_{j=1}^{J} \prod_{k=1}^{K} \frac{\lambda_{jk}^{y_{jk}} e^{-\lambda_{jk}}}{y_{jk}!} = n! \prod_{j=1}^{J} \prod_{k=1}^{K} \frac{\theta_{jk}^{y_{jk}}}{y_{jk}!}$$

where $\theta_{jk} = \lambda_{jk}/\lambda$. This is the *multinomial distribution.* It provides a suitable

Table 9.6 Notation for two-dimensional tables.

	B_1	B_2	...	B_K	Total
A_1	Y_{11}	Y_{12}	...	Y_{1K}	$Y_{1.}$
A_2	Y_{21}				$Y_{2.}$
$\vdots$	$\vdots$				
A_J	Y_{J1}			Y_{JK}	$Y_{J.}$
Total	$Y_{.1}$	$Y_{.2}$		$Y_{.K}$	$n = Y_{..}$

model for the tumour data in Example 9.1. By definition $0 \leqslant \theta_{jk} \leqslant 1$ and $\sum_j \sum_k \theta_{jk} = 1$; thus the terms θ_{jk} represent the probabilities of the cells.

For tables in which the row (column) totals are fixed, the probability distribution for each row (column) is multinomial and the rows (columns) are assumed to be independent. Thus for fixed row totals the joint probability distribution of the Y_{jk}s is

$$f(\mathbf{y} | y_{j.}, j = 1, \ldots, J) = \prod_{j=1}^{J} y_{j.}! \prod_{k=1}^{K} \frac{\theta_{jk}^{y_{jk}}}{y_{jk}!}$$

with $\sum_k \theta_{jk} = 1$ for each j. This is the *product multinomial distribution* and it is a suitable model for the randomized controlled trial data in Example 9.2.

For contingency tables with more than two dimensions the three major probability distributions are as follows.

Poisson distribution

$$f(\mathbf{y}; \boldsymbol{\lambda}) = \prod_{i=1}^{N} \lambda_i^{y_i} e^{-\lambda_i} / y_i! \tag{9.1}$$

with no constraints on the frequencies y_i or on the parameters λ_i.

Multinomial distribution

$$f(\mathbf{y}; \boldsymbol{\theta} | n) = n! \prod_{i=1}^{N} \theta_i^{y_i} / y_i! \tag{9.2}$$

where $n = \sum_{i=1}^{N} y_i$ and $\sum_{i=1}^{N} \theta_i = 1$.

Product multinomial distribution

For a three-dimensional table with J rows, K columns and L layers (subtables), if the row totals are fixed in each layer

$$f(\mathbf{y}; \boldsymbol{\theta} | y_{j.l}, j = 1, \ldots, J; l = 1, \ldots, L) = \prod_{j=1}^{J} \prod_{l=1}^{L} y_{j.l}! \prod_{k=1}^{K} \theta_{jkl}^{y_{jkl}} / y_{jkl}! \tag{9.3}$$

with $y_{j.l} = \sum_k y_{jkl}$ fixed and $\sum_k \theta_{jkl} = 1$ for $j = 1, \ldots, J$ and $l = 1, \ldots, L$. If only the layer totals are fixed the distribution is

$$f(\mathbf{y};\boldsymbol{\theta}|y_{..l}, l = 1, \ldots, L) = \prod_{l=1}^{L} y_{..l}! \prod_{j=1}^{J} \prod_{k=1}^{K} \theta_{jkl}^{y_{jkl}}/y_{jkl}! \tag{9.4}$$

with $y_{..l} = \sum_j \sum_k y_{jkl}$ fixed and $\sum_j \sum_k \theta_{jkl} = 1$ for each l.

The distribution given in (9.3) is a suitable model for the ulcer and blood group data (Example 9.3) with $J = 2$ for cases or controls, $K = 2$ blood groups and $L = 3$ cities.

9.3 Log-linear models

For any multinomial distribution with cell frequencies $Y_1, \ldots, Y_N$, cell probabilities $\theta_1, \ldots, \theta_N$ with $\sum_{i=1}^{N} \theta_i = 1$ and total frequency $n = \sum_{i=1}^{N} Y_i$, it can be shown that

$$E(Y_i) = n\theta_i, \quad i = 1, \ldots, N \tag{9.5}$$

(for example, see Bishop *et al.*, 1975, Section 13.4). Hence for the product multinomial distributions in (9.3) and (9.4)

$$E(Y_{jkl}) = y_{j.l}\,\theta_{jkl} \tag{9.6}$$

and

$$E(Y_{jkl}) = y_{..l}\,\theta_{jkl},$$

respectively. For the Poisson distribution (9.1) the expected cell frequencies are given by $E(Y_i) = \lambda_i$.

For the two-dimensional contingency tables the usual hypotheses can all be formulated as multiplicative models for the expected cell frequencies. For example, for the multinomial distribution if the row and column variables are *independent* then $\theta_{jk} = \theta_{j.}\,\theta_{.k}$ so

$$E(Y_{jk}) = n\theta_{j.}\theta_{.k}, \tag{9.7}$$

where $\theta_{j.}$ and $\theta_{.k}$ represent probabilities of levels of row and column variables, respectively, and $\sum_j \theta_{j.} = \sum_k \theta_{.k} = 1$.

For the product multinomial distribution with fixed row totals $y_{j.}$, the hypothesis that the cell probabilities are the same across all columns, the *homogeneity* hypothesis, can be represented as

$$E(Y_{jk}) = y_{j.}\theta_{.k}. \quad \text{for} \quad j = 1, \ldots, J \quad \text{and} \quad k = 1, \ldots, K$$

with $\sum_k \theta_{.k} = 1$.

Similarly for tables in higher dimensions the most common hypotheses are expressed as multiplicative models for expected cell frequencies.

This suggests that for generalized linear models the logarithm is the natural link function between $E(Y_i)$ and a linear combination of parameters, i.e.

$$\eta_i = \log E(Y_i) = \mathbf{x}_i^{\mathrm{T}} \boldsymbol{\beta}, \quad i = 1, \ldots, N,$$

hence the name *log-linear model*. For example, (9.7) can be expressed as

$$\eta_{jk} = \log E(Y_{jk}) = \mu + \alpha_j + \beta_k \tag{9.8}$$

and, by analogy with analysis of variance, the corresponding maximal model $E(Y_{jk}) = n\theta_{jk}$ can be written as

$$\eta_{jk} = \log E(Y_{jk}) = \mu + \alpha_j + \beta_k + (\alpha\beta)_{jk} \tag{9.9}$$

so that the independence hypothesis $\theta_{jk} = \theta_{j.}\theta_{.k}$ for all j and k is equivalent to the 'no interaction' hypothesis that $(\alpha\beta)_{jk} = 0$ for all j and k.

Log-linear models are usually *hierarchical* in the sense that higher order terms are defined as deviations from lower order terms. For example, in (9.8) α_j represents the differential effect of row j beyond the average effect μ. Generally, higher order terms are not included in a model unless all the related lower order terms are included.

As with ANOVA models the log-linear models (9.8) and (9.9) have too many parameters so that sum-to-zero or corner-point constraints are needed. In general, for *main effects* α_j where $j = 1, \ldots, J$ there are $(J-1)$ independent parameters; for *first order interactions* $(\alpha\beta)_{jk}$ where $j = 1, \ldots, J$ and $k = 1, \ldots, K$, there are $(J-1)(K-1)$ independent parameters, and so on.

In the expressions for expected cell frequencies for multinomial and product multinomial distributions certain terms are fixed constants, for instance n in (9.5) or $y_{j.l}$ in (9.6). This means that the corresponding parameters must always be included in the log-linear models. For example, the maximal model corresponding to $E(Y_{jkl}) = y_{j.l}\theta_{jkl}$ in (9.6) is

$$\eta_{jkl} = \mu + \alpha_j + \beta_k + \gamma_l + (\alpha\beta)_{jk} + (\alpha\gamma)_{jl} + (\beta\gamma)_{kl} + (\alpha\beta\gamma)_{jkl}$$

(with the parameters subject to appropriate constraints) in which the expression

$$\mu + \alpha_j + \gamma_l + (\alpha\gamma)_{jl} \tag{9.10}$$

corresponds to the fixed marginal total $y_{j.l}$ and the remainder

$$\beta_k + (\alpha\beta)_{jk} + (\beta\gamma)_{kj} + (\alpha\beta\gamma)_{jkl} \tag{9.11}$$

to the cell probability θ_{jkl}. Thus any hypothesis about the structure of the cell probabilities is formulated by omitting terms from (9.11) while the expression (9.10) is a necessary part of any model.

Table 9.7 summarizes the most commonly used log-linear models for two-dimensional contingency tables. Generally the same models apply for all three probability distributions although there are differences in the terms which must be included in the models and differences in the interpretation

Table 9.7 Log-linear models for two-dimensional contingency tables.

Log-linear model	Poisson distribution	Multinomial distribution	Product multinomial distribution with $y_{j.}$ fixed
Maximal model $\mu+\alpha_j+\beta_k+(\alpha\beta)_{jk}$ with JK independent parameters	$E(Y_{jk})=\lambda_{jk}$	$E(Y_{jk})=n\theta_{jk}$ with $\sum_j\sum_k\theta_{jk}=1$	$E(Y_{jk})=y_{j.}\theta_{jk}$ with $\sum_k\theta_{jk}=1$ for $j=1,\ldots,J$.
$\mu+\alpha_j+\beta_k$ with $J+K-1$ independent parameters	independence hypothesis $E(Y_{jk})=\lambda'_j\lambda''_k$	independence hypothesis $E(Y_{jk})=n\theta_{j.}\theta_{.k}$ with $\sum_j\theta_{j.}=\sum_k\theta_{.k}=1$	homogeneity hypothesis $E(Y_{jk})=y_{j.}\theta_{.k}$ with $\sum_k\theta_{.k}=1$
Terms which must be included in any log-linear model		μ since n is fixed	$\mu+\alpha_j$ since $y_{j.}$ is fixed

of the 'interaction' term. Three-dimensional tables models corresponding to the major hypotheses for multinomial and product multinomial distributions are given in Appendix 4.

9.4 Maximum likelihood estimation

For the Poisson distribution (9.1) the log-likelihood function is

$$l=\sum_i(y_i\log\lambda_i-\lambda_i-\log y_i!)$$

where $\lambda_i=E(Y_i)$. Hence for the maximal model the maximum likelihood estimator of λ_i is the solution of $\partial l/\partial\lambda_i=(y_i/\lambda_i)-1$, i.e. $\hat{\lambda}_i=Y_i$.

For the multinomial distribution (9.2) the log-likelihood function is

$$l=\log n!+\sum_i(y_i\log\theta_i-\log y_i!)$$

which can be written in the form

$$l=\text{constant}+\sum_i y_i\log\theta_i$$

where $n\theta_i=E(Y_i)$. For the maximal model, the maximum likelihood estimators of the θ_is are obtained by maximizing l subject to the constraint that $\sum\theta_i=1$. This can be done using a *Lagrange multiplier*, that is finding ξ and θ_i to minimize

$$t=\text{constant}+\sum_i y_i\log\theta_i-\xi(\sum\theta_i-1).$$

The solutions of $\partial t/\partial\xi = 0$ and $\partial t/\partial\theta_i = 0$ for $i = 1, \ldots, N$ are $\xi = n$ and $\hat{\theta}_i = Y_i/n$. So the maximum likelihood estimator of $E(Y_i)$ is $n\hat{\theta}_i = Y_i$.

Similarly for any product multinomial distribution the log-likelihood function can be written in the form

$$l = \text{constant} + \sum_i y_i \log E(Y_i),$$

and so for the maximal model the maximum likelihood estimator of $E(Y_i)$ is Y_i.

Thus for all three probability distributions the log-likelihood function depends only on the observed cell frequencies $\mathbf{y}$ and their expected frequencies $E(\mathbf{y})$ where $\boldsymbol{\eta} = \log E(\mathbf{y}) = \mathbf{X}\boldsymbol{\beta}$. If we are mainly interested in hypothesis testing we can find the maximum likelihood estimates for $E(\mathbf{y})$ directly and then calculate the log-likelihood ratio statistic. On the other hand, if we are interested in the parameters of the linear component $\mathbf{X}\boldsymbol{\beta}$, we can begin by estimating these. The choice of which parameters, $E(\mathbf{y})$ or $\boldsymbol{\beta}$, to estimate depends on the application and computational convenience. By the invariance property of maximum likelihood estimators, either set of estimators can be readily obtained from the other.

Birch (1963) showed that for any log-linear model the maximum likelihood estimators are the same for all three probability distributions, provided that the parameters which correspond to the fixed marginal totals are always included in the model (as discussed in Section 9.3). Thus for the purpose of estimation the Poisson distribution can be assumed. Since this involves unrestricted independent random variables with distributions from the exponential family, the Newton–Raphson type estimation procedures described in Chapter 5 may be used to estimate $\boldsymbol{\beta}$ and hence $E(\mathbf{y})$. This approach has been advocated by Nelder (1974) and is implemented in GLIM.

An alternative approach, based on estimating the expected cell frequencies $E(\mathbf{y})$, is to obtain analytic expressions for the maximum likelihood estimators often using Lagrange multipliers to incorporate any constraints on the probabilities. For the expected cell frequencies explicit closed-form solutions of the maximum likelihood equations may not exist so approximate numerical solutions have to be calculated. An iterative method is used to adjust the estimated expected cell frequencies until they add up to the required marginal totals (at least to within some specified accuracy). This procedure is called *iterative proportional fitting*. It is described in detail by Bishop *et al.* (1975) and it is implemented in most statistical computer packages.

Maximum likelihood estimation is illustrated by numerical examples in Section 9.6 after hypothesis testing has been considered.

9.5 Hypothesis testing and goodness of fit

The statistics used to measure goodness of fit for contingency tables are closely related to those for binary data considered in the previous chapter.

With any of the three probability distributions, the estimated expected cell frequencies for the maximal model are the observed cell frequencies y_i so the log-likelihood function has the form

$$l(\mathbf{b}_{\max}; \mathbf{y}) = \text{constant} + \sum_i y_i \log y_i.$$

For any other model let e_i denote the estimated expected cell frequencies so that the log-likelihood function is

$$l(\mathbf{b}: \mathbf{y}) = \text{constant} + \sum_i y_i \log e_i.$$

Hence the log-likelihood ratio statistic is

$$D = 2[l(\mathbf{b}_{\max}; \mathbf{y}) - l(\mathbf{b}; \mathbf{y})] = 2\sum_{i=1}^{N} y_i \log \frac{y_i}{e_i}$$

which is of the form

$$D = 2\sum o \log \frac{o}{e},$$

where o and e denote the observed and estimated expected (i.e. fitted) cell frequencies, respectively, and summation is over all cells in the table. From Chapter 5, if the model fits the data well for large samples, then D has the central chi-squared distribution with *degrees of freedom* given by the *number of cells with non-zero observed frequencies* (i.e. N if $y_i > 0$ for all i) *minus the number of independent, non-zero parameters* in the model.

The chi-squared statistic

$$X^2 = \sum \frac{(o-e)^2}{e}$$

is more commonly used for contingency table data than D. By the argument used in Section 8.5, it can readily be shown that these two statistics are asymptotically equivalent and hence that, for large samples, X^2 has the chi-squared distribution with the number of degrees of freedom given above.

The form of the chi-squared statistic, and indeed the Poisson model in which $E(o) = \text{var}(o) = e$, suggest that the standardized residual for each cell be defined as

$$r_i = (o_i - e_i)/e_i^{\frac{1}{2}}.$$

Departures from the model may be detected by inspecting the residuals. Values which are too far from zero in either direction (say $|r_i| > 3$ corresponding roughly to the 1% tails of the standard Normal distribution) or patterns in the residuals from certain parts of the table may suggest more appropriate models.

9.6 Numerical examples

(i) *Tumour type and site* (Example 9.1)

We use the multinomial distribution (9.2). Under the independence hypothesis

$$H_0: E(Y_{jk}) = n\theta_{j.}\theta_{.k}.$$

The maximum likelihood estimators of $\theta_{j.}$ and $\theta_{.k}$ with

$$\sum_j \theta_{j.} = \sum_k \theta_{.k} = 1$$

can be obtained by maximizing

$$t = \text{constant} + \sum_j \sum_k y_{jk} \log(\theta_{j.}\theta_{.k}) - \xi_1\Big(\sum_j \theta_{j.} - 1\Big) - \xi_2\Big(\sum_k \theta_{.k} - 1\Big).$$

The solutions of

$$\frac{\partial t}{\partial \theta_{j.}} = \frac{\partial t}{\partial \theta_{.k}} = \frac{\partial t}{\partial \xi_1} = \frac{\partial t}{\partial \xi_2} = 0$$

give $\hat{\theta}_{j.} = y_{j.}/n$ and $\hat{\theta}_{.k} = y_{.k}/n$ so that the fitted cell frequencies are

$$e_{jk} = \frac{y_{j.}y_{.k}}{n}, \quad j = 1, \ldots, J; k = 1, \ldots, K.$$

The terms of the log-linear model under H_0 are

$$\eta_{jk} = \log E(Y_{jk}) = \mu + \alpha_j + \beta_k$$

with $\sum_j \alpha_j = \sum_k \beta_k = 0$. From the estimate $\log e_{jk} = \hat{\eta}_{jk}$

$$\log y_{j.} + \log y_{.k} - \log n = \hat{\mu} + \hat{\alpha}_j + \hat{\beta}_k.$$

We add these over j and k to obtain the minimum variance estimates

$$\hat{\mu} = \frac{1}{J}\sum_j \log y_{j.} + \frac{1}{K}\sum_k \log y_{.k} - \log n, \quad \hat{\alpha}_j = \log y_{j.} - \frac{1}{J}\sum_j \log y_{j.}$$

and

$$\hat{\beta}_k = \log y_{.k} - \frac{1}{K}\sum_k \log y_{.k}.$$

Under the maximal model, from Section 9.4, the estimated expected cell frequencies are the observed frequencies y_{jk} so

$$\hat{\eta}_{jk} = \hat{\mu} + \hat{\alpha}_j + \hat{\beta}_k + (\widehat{\alpha\beta})_{jk} = \log y_{jk}.$$

Hence

$$\hat{\mu} = \frac{1}{JK}\sum_j \sum_k \log y_{jk}, \quad \hat{\alpha}_j = \frac{1}{K}\sum_k \log y_{jk} - \hat{\mu}, \quad \hat{\beta}_k = \frac{1}{J}\sum_j \log y_{jk} - \hat{\mu}$$

and

$$\widehat{(\alpha\beta)}_{jk} = \log y_{jk} - \hat{\mu} - \hat{\alpha}_j - \hat{\beta}_k.$$

The numerical results for the tumour data are shown in Table 9.8. The model corresponding to H_0 fits the data poorly so we reject the independence hypothesis. The largest residual, for cell (1, 1), accounts for much of the lack of fit confirming that the main 'signal' in the data is the association of Hutchinson's melanotic freckle with the head and neck.

Table 9.8(a) Analysis of tumour type and site data.

Cell j	Cell k	Observed frequency	Expected frequency under H_0	Standardized residual
1	1	22	5.780	6.75
1	2	2	9.010	−2.34
1	3	10	19.210	−2.10
2	1	16	31.450	−2.75
2	2	54	49.025	0.71
2	3	115	104.525	1.02
3	1	19	21.250	−0.49
3	2	33	33.125	−0.02
3	3	73	70.625	0.28
4	1	11	9.520	0.48
4	2	17	14.840	0.56
4	3	28	31.640	−0.65

(*b*) Estimates of log-linear main effect parameters.

Parameter	Maximal model	Model under H_0
μ	3.074	3.176
α_1	−1.045	−0.874
α_2	0.762	0.820
α_3	0.503	0.428
α_4	−0.220	−0.375
β_1	−0.273	−0.548
β_2	−0.321	−0.104
β_3	0.594	0.653

Goodness of fit under H_0: $D = 51.8$, $X^2 = 65.8$, $\Pr(\chi^2_6 > 50) < 0.001$

(ii) *Peptic ulcers and blood groups* (Example 9.3)

We use the product multinomial distribution, (9.3), and test the hypothesis that the relationship between blood groups and cases or controls is the same in all cities, i.e. $E(Y_{jkl}) = y_{j.l}\theta_{jk.}$, or equivalently

$$\eta_{jkl} = \log E(Y_{jkl}) = \mu+\alpha_j+\gamma_l+(\alpha\gamma)_{jl}+\beta_k+(\alpha\beta)_{jk} \qquad (9.12)$$

where the first four terms correspond to the fixed marginal totals $y_{j.l}$. Maximizing the log-likelihood function subject to the constraints $\sum_k \theta_{jk.} = 1$ for each j gives the fitted cell frequencies

$$e_{jk} = y_{j.l}y_{jk.}/y_{j..}.$$

The number of independent parameters in model (9.12) is

$$1+(J-1)+(L-1)+(J-1)(L-1)+(K-1)+(J-1)(K-1) = J(K+L-1)$$

so the appropriate degrees of freedom are

$$JKL-J(K+L-1) = J(K-1)(L-1).$$

The results for these data are given in Table 9.9. The model provides a poor description of the data with the largest residuals (marked*) corresponding to the control subjects from London. Thus we would reject the hypothesis that the relationship between blood groups and cases or controls is the same for all cities. (Further analyses of these data are suggested in the exercises.)

Table 9.9 Analysis of peptic ulcer and blood group data.

Cell			Observed frequency	Expected frequency	Standardized residual
j	k	l			
1	1	1	579	597.3	−0.75
2	1	1	4219	4026.0	3.04*
1	2	1	911	892.7	0.61
2	2	1	4578	4771.0	−2.79*
1	1	2	246	243.3	0.17
2	1	2	3775	3801.7	−0.43
1	2	2	361	363.7	−0.14
2	2	2	4532	4505.3	0.40
1	1	3	291	275.4	0.94
2	1	3	5261	5427.3	−2.26
1	2	3	396	411.6	−0.77
2	2	3	6598	6431.7	2.07

$D = 29.24$, $X^2 = 29.26$, $\Pr(\chi^2_4 > 29) > 0.001$

9.7 Remarks

The numerical examples in the previous section are particularly simple in that the calculations do not require iteration and the interpretation of the results is straightforward. For contingency tables involving more than three variables, model selection and interpretation become much more complicated. Some suggestions for systematically fitting complex log-linear models are given by Bishop *et al.* (1975), Chs. 4 and 9, and Whittaker and Aitkin (1978). The analysis of multidimensional contingency tables usually requires a computer to perform the iterative estimation.

An alternative approach to the likelihood methods considered in this chapter has been proposed by Grizzle *et al.* (1969). It is based on modelling functions of the multinomial probabilities $\boldsymbol{\theta}$ as linear combinations of parameters, i.e.

$$\mathbf{F}(\boldsymbol{\theta}) = \mathbf{X}\boldsymbol{\beta}$$

and using the weighted least squares criterion

$$S_w = [\mathbf{F}(\mathbf{p}) - \mathbf{X}\boldsymbol{\beta}]^{\mathrm{T}} \mathbf{V}^{-1} [\mathbf{F}(\mathbf{p}) - \mathbf{X}\boldsymbol{\beta}]$$

for estimation and hypothesis testing (where $\mathbf{p}$ represents the estimated probabilities and $\mathbf{V}$ the variance–covariance matrix for $\mathbf{F}(\mathbf{p})$). An advantage of this method is that it can be used for linear and non-linear (including log-linear) models. But it is computationally more complex than the likelihood methods and is less widely used.

Contingency table methods, including log-linear models, are primarily designed for analysing data for nominal categories. In practice they are also used for ordinal categories, either ignoring the ordering or assigning covariate scores to the categories. Recently McCullagh (1980) has shown that generalized linear modelling can be extended to give regression-like methods for ordinal data. The details are beyond the scope of this book and the reader is referred to the original paper or McCullagh and Nelder (1983).

9.8 Exercises

9.1 For the randomized controlled trial, Example 9.2:

(i) derive maximum likelihood estimators for the parameters in the log-linear model and the expected cell frequencies under

(a) the maximal model,

(b) the model corresponding to the homogeneity hypothesis;

(ii) verify that the estimators are the same as those for the multinomial distribution (see tumour example in Section 9.6(i));

(iii) test the hypothesis that there is no difference between the treatments.

9.2. Test the hypotheses that the distributions of blood groups are the same for peptic ulcer cases and controls in Example 9.3 in both Manchester and Newcastle upon Tyne (since the data for London are not comparable, see Section 9.6(ii)).

9.3 For a 2×2 contingency table the maximal log-linear model can be written as

$$\eta_{11} = \mu+\alpha+\beta+(\alpha\beta), \quad \eta_{12} = \mu+\alpha-\beta-(\alpha\beta),$$
$$\eta_{21} = \mu-\alpha+\beta-(\alpha\beta), \quad \eta_{22} = \mu-\alpha-\beta+(\alpha\beta).$$

where $\eta_{jk} = \log E(Y_{jk}) = \log(n\theta_{jk})$ and $n = \Sigma\Sigma\, Y_{jk}$.
Show that the 'interaction' term $(\alpha\beta)$ is given by

$$(\alpha\beta) = \tfrac{1}{4}\log\psi$$

where ψ is the odds ratio $(\theta_{11}\theta_{22})/(\theta_{12}\theta_{21})$ (so that $(\alpha\beta) = 0$ is equivalent to $\psi = 1$).

9.4 Consider a $2 \times I$ contingency table in which the column totals n_i are fixed for $i = 1, \ldots, I$.

	1	...	i	...	I
Success	R_1		R_i		R_I
Failure	$n_1 - R_1$		$n_i - R_i$		$n_I - R_I$
Total	n_1		n_i		n_I

For the product binomial distribution $\theta_{1i} = \pi_i$ and $\theta_{2i} = 1-\pi_i$.

(i) Show that the log-linear model with

$$\eta_{1i} = \log E(R_i) = \mathbf{x}_{1i}^{\mathrm{T}}\boldsymbol{\beta}$$

and

$$\eta_{2i} = \log E(n_i - R_i) = \mathbf{x}_{2i}^{\mathrm{T}}\boldsymbol{\beta}$$

is equivalent to the logistic model

$$\log\left(\frac{\pi_i}{1-\pi_i}\right) = \mathbf{x}_i^{\mathrm{T}}\boldsymbol{\beta}$$

where $\mathbf{x}_i = \mathbf{x}_{1i} - \mathbf{x}_{2i}$.

(ii) Analyse the peptic ulcer data (for Manchester and Newcastle upon Tyne only) using logistic regression and compare the results from those obtained in Exercise 9.2.

APPENDIX 1

Consider a continuous random variable Y with probability density function $f(y;\theta)$ depending on a single parameter θ (or if Y is discrete $f(y;\theta)$ is its probability distribution). The log-likelihood function is the logarithm of $f(y;\theta)$ regarded primarily as a function of θ, i.e.

$$l(\theta;y) = \log f(y;\theta).$$

The derivative $U = \mathrm{d}l/\mathrm{d}\theta$ is called the *score*. To find the moments of U we use the identity

$$\frac{\mathrm{d}\log f(y;\theta)}{\mathrm{d}\theta} = \frac{1}{f(y;\theta)}\frac{\mathrm{d}f(y;\theta)}{\mathrm{d}\theta}. \tag{A1.1}$$

If we take expectations of (A1.1) we obtain

$$E(U) = \int \frac{\mathrm{d}\log f(y;\theta)}{\mathrm{d}\theta} f(y;\theta)\,\mathrm{d}y = \int \frac{\mathrm{d}f(y;\theta)}{\mathrm{d}\theta}\,\mathrm{d}y$$

where integration is over the domain of y. Under certain regularity conditions the right-hand term is

$$\int \frac{\mathrm{d}f(y;\theta)}{\mathrm{d}\theta}\,\mathrm{d}y = \frac{\mathrm{d}}{\mathrm{d}\theta}\int f(y;\theta)\,\mathrm{d}y = \frac{\mathrm{d}}{\mathrm{d}\theta}1 = 0$$

since $\int f(y;\theta)\,\mathrm{d}y = 1$. Hence

$$E(U) = 0 \tag{A1.2}$$

Also if we differentiate (A1.1) with respect to θ and take expectations, provided the order of these operations can be interchanged, then

$$\frac{\mathrm{d}}{\mathrm{d}\theta}\int \frac{\mathrm{d}\log f(y;\theta)}{\mathrm{d}\theta} f(y;\theta)\,\mathrm{d}y = \frac{\mathrm{d}^2}{\mathrm{d}\theta^2}\int f(y;\theta)\,\mathrm{d}y.$$

The right-hand side equals zero because $\int f(y;\theta)\,\mathrm{d}y = 1$ and the left-hand side can be expressed as

$$\int \frac{\mathrm{d}^2 \log f(y;\theta)}{\mathrm{d}\theta^2} f(y;\theta)\,\mathrm{d}y + \int \frac{\mathrm{d}\log f(y;\theta)}{\mathrm{d}\theta}\frac{\mathrm{d}f(y;\theta)}{\mathrm{d}\theta}\,\mathrm{d}y.$$

Hence, substituting (A1.1) in the second term, we obtain

$$\int \frac{\mathrm{d}^2 \log f(y;\theta)}{\mathrm{d}\theta^2} f(y;\theta)\,\mathrm{d}y + \int \left[\frac{\mathrm{d}\log f(y;\theta)}{\mathrm{d}\theta}\right]^2 f(y;\theta)\,\mathrm{d}y = 0.$$

Therefore,

$$E\left[-\frac{\mathrm{d}^2 \log f(y;\theta)}{\mathrm{d}\theta^2}\right] = E\left\{\left[\frac{\mathrm{d}\log f(y;\theta)}{\mathrm{d}\theta}\right]^2\right\},$$

i.e.

$$E(-U') = E(U^2).$$

Since $E(U) = 0$ the variance of U, which is called the *information*, is

$$\operatorname{var}(U) = E(U^2) = E(-U'). \tag{A1.3}$$

More generally consider independent random variables $Y_1, \ldots, Y_N$ whose probability distributions depend on parameters $\theta_1, \ldots, \theta_p$ where $p \leqslant N$. Let $l_i(\boldsymbol{\theta}; y_i)$ denote the log-likelihood function of Y_i where $\boldsymbol{\theta} = [\theta_1, \ldots, \theta_p]^{\mathrm{T}}$. Then the log-likelihood function of $Y_1, \ldots, Y_N$ is

$$l(\boldsymbol{\theta}; \mathbf{y}) = \sum_{i=1}^{N} l_i(\boldsymbol{\theta}; y_i)$$

where $\mathbf{y} = [y_1, \ldots, y_N]^{\mathrm{T}}$. The *total score* with respect to θ_j is defined as

$$U_j = \frac{\partial l(\boldsymbol{\theta}; \mathbf{y})}{\partial \theta_j} = \sum_{i=1}^{N} \frac{\partial l_i(\boldsymbol{\theta}; y_i)}{\partial \theta_j}.$$

By the same argument as for (A1.2),

$$E\left[\frac{\partial l_i(\boldsymbol{\theta}; y_i)}{\partial \theta_j}\right] = 0$$

and so

$$E(U_j) = 0 \quad \text{for all } j. \tag{A1.4}$$

The *information matrix* is defined to be the variance–covariance matrix of the U_js, $\mathscr{I} = E(\mathbf{U}\mathbf{U}^{\mathrm{T}})$ where $\mathbf{U} = [U_1, \ldots, U_p]^{\mathrm{T}}$, so it has elements

$$\mathscr{I}_{jk} = E\,[U_j U_k] = E\left[\frac{\partial l}{\partial \theta_j}\frac{\partial l}{\partial \theta_k}\right]. \tag{A1.5}$$

By an argument analogous to the single random variable, single parameter case above, it can be shown that

$$E\left[\frac{\partial l_i}{\partial \theta_j}\frac{\partial l_i}{\partial \theta_k}\right] = E\left[-\frac{\partial^2 l_i}{\partial \theta_j \partial \theta_k}\right].$$

Hence the elements of the information matrix are also given by

$$\mathscr{I}_{jk} = E\left[-\frac{\partial^2 l}{\partial \theta_j \partial \theta_k}\right]. \qquad \text{(A1.6)}$$

APPENDIX 2

From sections 3.2 and 3.3, for the generalized linear model the log-likelihood function can be written as

$$l(\boldsymbol{\theta}; \mathbf{y}) = \Sigma\, y_i b(\theta_i) + \Sigma\, c(\theta_i) + \Sigma\, d(y_i)$$

with

$$E(Y_i) = \mu_i = -c'(\theta_i)/b'(\theta_i) \qquad \text{(A2.1)}$$

and

$$g(\mu_i) = \mathbf{x}_i^{\mathrm{T}}\boldsymbol{\beta} = \sum_{j=1}^{p} x_{ij}\beta_j = \eta_i \qquad \text{(A2.2)}$$

where g is a monotone, differentiable function. Also from (3.5)

$$\operatorname{var}(Y_i) = [b''(\theta_i)c'(\theta_i) - c''(\theta_i)b'(\theta_i)]/[b'(\theta_i)]^3 \qquad \text{(A2.3)}$$

The score with respect to parameter β_j is defined as

$$U_j = \frac{\partial l(\boldsymbol{\theta}; \mathbf{y})}{\partial \beta_j} = \sum_{i=1}^{N} \frac{\partial l_i}{\partial \beta_j}$$

where

$$l_i = y_i b(\theta_i) + c(\theta_i) + d(y_i). \qquad \text{(A2.4)}$$

To obtain U_j we use

$$\frac{\partial l_i}{\partial \beta_j} = \frac{\partial l_i}{\partial \theta_i}\frac{\partial \theta_i}{\partial \mu_i}\frac{\partial \mu_i}{\partial \beta_j}.$$

By differentiating (A2.4) and substituting (A2.1)

$$\frac{\partial l_i}{\partial \theta_i} = y_i b'(\theta_i) + c'(\theta_i) = b'(\theta_i)(y_i - \mu_i).$$

By differentiating (A2.1) and substituting (A2.3)

$$\frac{\partial \mu_i}{\partial \theta_i} = -\frac{c''(\theta_i)}{b'(\theta_i)} + \frac{c'(\theta_i)b''(\theta_i)}{[b'(\theta_i)]^2} = b'(\theta_i)\operatorname{var}(Y_i)$$

By differentiating (A2.2)

$$\frac{\partial \mu_i}{\partial \beta_j} = \frac{\partial \mu_i}{\partial \eta_i}\frac{\partial \eta_i}{\partial \beta_j} = x_{ij}\frac{\partial \mu_i}{\partial \eta_i}.$$

Hence

$$\frac{\partial l_i}{\partial \beta_j} = \frac{\partial l_i}{\partial \theta_i}\frac{\partial \mu_i}{\partial \beta_j}\Big/\frac{\partial \mu_i}{\partial \theta_i} = \frac{(y_i - \mu_i)x_{ij}}{\text{var}(Y_i)}\left(\frac{\partial \mu_i}{\partial \eta_i}\right) \tag{A2.5}$$

and therefore

$$U_j = \sum_{i=1}^{N} = \frac{(y_i - \mu_i)x_{ij}}{\text{var}(Y_i)}\left(\frac{\partial \mu_i}{\partial \eta_i}\right). \tag{A2.6}$$

The elements of the information matrix are defined by $\mathscr{I}_{jk} = E(U_j U_k)$. From (A2.5), for each Y_i the contribution to $\mathscr{I}_{jk}$ is

$$E\left[\frac{\partial l_i}{\partial \beta_j}\frac{\partial l_i}{\partial \beta_k}\right] = E\left[\frac{(y_i - \mu_i)^2 x_{ij} x_{ik}}{\{\text{var}(Y_i)\}^2}\left(\frac{\partial \mu_i}{\partial \eta_i}\right)^2\right] = \frac{x_{ij} x_{ik}}{\text{var}(Y_i)}\left(\frac{\partial \mu_i}{\partial \eta_i}\right)^2$$

and therefore

$$\mathscr{I}_{jk} = \sum_{i=1}^{N} \frac{x_{ij} x_{ik}}{\text{var}(Y_i)}\left(\frac{\partial \mu_i}{\partial \eta_i}\right)^2. \tag{A2.7}$$

APPENDIX 3

Here are several versions of analysis of variance for the two factor experiment shown in Table 7.5. The responses are

$$\mathbf{y} = [6.8, 6.6, 5.3, 6.1, 7.5, 7.4, 7.2, 6.5, 7.8, 9.1, 8.8, 9.1]^{\mathrm{T}}$$

A3.1 Conventional parameterizations with sum-to-zero constraints

(A) FULL MODEL: $E(Y_{jkl}) = \mu + \alpha_j + \beta_k + (\alpha\beta)_{jk}$

$$\boldsymbol{\beta} = \begin{bmatrix} \mu \\ \alpha_1 \\ \alpha_2 \\ \alpha_3 \\ \beta_1 \\ \beta_2 \\ (\alpha\beta)_{11} \\ (\alpha\beta)_{12} \\ (\alpha\beta)_{21} \\ (\alpha\beta)_{22} \\ (\alpha\beta)_{31} \\ (\alpha\beta)_{32} \end{bmatrix}, \quad \mathbf{X} = \begin{bmatrix} 1&1&0&0&1&0&1&0&0&0&0&0 \\ 1&1&0&0&1&0&1&0&0&0&0&0 \\ 1&1&0&0&0&1&0&1&0&0&0&0 \\ 1&1&0&0&0&1&0&1&0&0&0&0 \\ 1&0&1&0&1&0&0&0&1&0&0&0 \\ 1&0&1&0&1&0&0&0&1&0&0&0 \\ 1&0&1&0&0&1&0&0&0&1&0&0 \\ 1&0&1&0&0&1&0&0&0&1&0&0 \\ 1&0&0&1&1&0&0&0&0&0&1&0 \\ 1&0&0&1&1&0&0&0&0&0&1&0 \\ 1&0&0&1&0&1&0&0&0&0&0&1 \\ 1&0&0&1&0&1&0&0&0&0&0&1 \end{bmatrix}, \quad \mathbf{X}^{\mathrm{T}}\mathbf{y} = \begin{bmatrix} Y_{...} \\ Y_{1..} \\ Y_{2..} \\ Y_{3..} \\ Y_{.1.} \\ Y_{.2.} \\ Y_{11.} \\ Y_{12.} \\ Y_{21.} \\ Y_{22.} \\ Y_{31.} \\ Y_{32.} \end{bmatrix} \begin{bmatrix} 88.2 \\ 24.8 \\ 28.6 \\ 34.8 \\ 45.2 \\ 43.0 \\ 13.4 \\ 11.4 \\ 14.9 \\ 13.7 \\ 16.9 \\ 17.9 \end{bmatrix}$$

The 12×12 matrix $\mathbf{X}$ has rank 6 so we impose six extra conditions in order to solve the normal equations $\mathbf{X}^{\mathrm{T}}\mathbf{X}\mathbf{b} = \mathbf{X}^{\mathrm{T}}\mathbf{y}$. These conditions are:

$$\alpha_1 + \alpha_2 + \alpha_3 = 0, \quad \beta_1 + \beta_2 = 0,$$
$$(\alpha\beta)_{11} + (\alpha\beta)_{12} = 0, \quad (\alpha\beta)_{21} + (\alpha\beta)_{22} = 0,$$
$$(\alpha\beta)_{31} + (\alpha\beta)_{32} = 0, \quad \text{and} \quad (\alpha\beta)_{11} + (\alpha\beta)_{21} + (\alpha\beta)_{31} = 0.$$

Hence we obtain

$$\mathbf{b} = [7.35, \ -1.15, \ -0.2, \ 1.35, \ 0.1833, \ -0.1833, \ 0.3167, \ -0.3167, \ 0.1167, \ -0.1167, \ -0.4333, \ 0.4333]^{\mathrm{T}}.$$

and therefore $\mathbf{b}^{\mathrm{T}}\mathbf{X}^{\mathrm{T}}\mathbf{y} = 662.62$.

(B) ADDITIVE MODEL: $E(Y_{jkl}) = \mu + \alpha_j + \beta_k$

The design matrix is obtained by omitting the last six columns from matrix **X** in part (A) and

$$\boldsymbol{\beta} = \begin{bmatrix} \mu \\ \alpha_1 \\ \alpha_2 \\ \alpha_3 \\ \beta_1 \\ \beta_2 \end{bmatrix}, \quad \mathbf{X}^T\mathbf{X} = \begin{bmatrix} 12 & 4 & 4 & 4 & 6 & 6 \\ 4 & 4 & 0 & 0 & 2 & 2 \\ 4 & 0 & 4 & 0 & 2 & 2 \\ 4 & 0 & 0 & 4 & 2 & 2 \\ 6 & 2 & 2 & 2 & 6 & 0 \\ 6 & 2 & 2 & 2 & 0 & 6 \end{bmatrix}, \quad \mathbf{X}^T\mathbf{y} = \begin{bmatrix} 88.2 \\ 24.8 \\ 28.6 \\ 34.8 \\ 45.2 \\ 43.0 \end{bmatrix}$$

$\mathbf{X}^T\mathbf{X}$ has rank 4 so we impose the extra conditions $\alpha_1 + \alpha_2 + \alpha_3 = 0$ and $\beta_1 + \beta_2 = 0$ to obtain

$$\mathbf{b} = [7.35, \ -1.15, \ -0.2, \ 1.35, \ 0.1833, -0.1833]^T$$

and $\mathbf{b}^T\mathbf{X}^T\mathbf{y} = 661.4133$.

(C) MODEL OMITTING EFFECTS OF LEVELS OF B:
$E(Y_{jkl}) = \mu + \alpha_j$

The design matrix is obtained by omitting the last eight columns from matrix **X** in part (A) and

$$\boldsymbol{\beta} = \begin{bmatrix} \mu \\ \alpha_1 \\ \alpha_2 \\ \alpha_3 \end{bmatrix}, \quad \mathbf{X}^T\mathbf{X} = \begin{bmatrix} 12 & 4 & 4 & 4 \\ 4 & 4 & 0 & 0 \\ 4 & 0 & 4 & 0 \\ 4 & 0 & 0 & 4 \end{bmatrix}, \quad \mathbf{X}^T\mathbf{y} = \begin{bmatrix} 88.2 \\ 24.8 \\ 28.6 \\ 34.8 \end{bmatrix}$$

$\mathbf{X}^T\mathbf{X}$ has rank 3 so we impose the extra condition $\alpha_1 + \alpha_2 + \alpha_3 = 0$ to obtain $\mathbf{b} = [7.35, \ -1.15, \ -0.2, \ 1.35]^T$ and $\mathbf{b}^T\mathbf{X}^T\mathbf{y} = 661.01$.

(D) MODEL OMITTING EFFECTS OF LEVELS OF A:
$E(Y_{jkl}) = \mu + \beta_k$

The design matrix is given by columns 1, 5 and 6 of matrix **X** in part (A) and $\boldsymbol{\beta} = [\mu, \beta_1, \beta_2]^T$, $\mathbf{X}^T\mathbf{X}$ is 3×3 of rank 2 so we impose the constraint $\beta_1 + \beta_2 = 0$ to obtain $\mathbf{b} = [7.35, \ 0.1833, \ -0.1833]^T$ and $\mathbf{b}^T\mathbf{X}^T\mathbf{y} = 648.6733$.

(E) MODEL WITH ONLY A MEAN EFFECT: $E(Y_{jkl}) = \mu$

In this case $\mathbf{b} = [\hat{\mu}] = 7.35$ and $\mathbf{b}^T\mathbf{X}^T\mathbf{y} = 648.27$.

A3.2 Corner-point parameterizations

(A) FULL MODEL: $E(Y_{jkl}) = \mu + \alpha_j + \beta_k + (\alpha\beta)_{jk}$ with

$$\alpha_1 = \beta_1 = (\alpha\beta)_{11} = (\alpha\beta)_{12} = (\alpha\beta)_{21} = (\alpha\beta)_{31} = 0$$

$$\boldsymbol{\beta} = \begin{bmatrix} \mu \\ \alpha_2 \\ \alpha_3 \\ \beta_2 \\ (\alpha\beta)_{22} \\ (\alpha\beta)_{32} \end{bmatrix}, \quad \mathbf{X} = \begin{bmatrix} 1 & 0 & 0 & 0 & 0 & 0 \\ 1 & 0 & 0 & 0 & 0 & 0 \\ 1 & 0 & 0 & 1 & 0 & 0 \\ 1 & 0 & 0 & 1 & 0 & 0 \\ 1 & 1 & 0 & 0 & 0 & 0 \\ 1 & 1 & 0 & 0 & 0 & 0 \\ 1 & 1 & 0 & 1 & 1 & 0 \\ 1 & 1 & 0 & 1 & 1 & 0 \\ 1 & 0 & 1 & 0 & 0 & 0 \\ 1 & 0 & 1 & 0 & 0 & 0 \\ 1 & 0 & 1 & 1 & 0 & 1 \\ 1 & 0 & 1 & 1 & 0 & 1 \end{bmatrix}, \quad \mathbf{X}^T\mathbf{y} = \begin{bmatrix} Y_{...} \\ Y_{2..} \\ Y_{3..} \\ Y_{.2.} \\ Y_{22.} \\ Y_{32.} \end{bmatrix} = \begin{bmatrix} 88.2 \\ 28.6 \\ 34.8 \\ 43.0 \\ 13.7 \\ 17.9 \end{bmatrix}$$

$$\text{so } \mathbf{X}^T\mathbf{X} = \begin{bmatrix} 12 & 4 & 4 & 6 & 2 & 2 \\ 4 & 4 & 0 & 2 & 2 & 0 \\ 4 & 0 & 4 & 2 & 0 & 2 \\ 6 & 2 & 2 & 6 & 2 & 2 \\ 2 & 2 & 0 & 2 & 2 & 0 \\ 2 & 0 & 2 & 2 & 0 & 2 \end{bmatrix}, \quad \mathbf{b} = \begin{bmatrix} 6.7 \\ 0.75 \\ 1.75 \\ -1.0 \\ 0.4 \\ 1.5 \end{bmatrix}$$

and $\mathbf{b}\mathbf{X}^T\mathbf{y} = 662.62$

(B) ADDITIVE MODEL: $E(Y_{jkl}) = \mu + \alpha_j + \beta_k$ with $\alpha_1 = \beta_1 = 0$

The design matrix is obtained by omitting the last two columns of matrix **X** in part (A) and so

$$\boldsymbol{\beta} = \begin{bmatrix} \mu \\ \alpha_2 \\ \alpha_3 \\ \beta_2 \end{bmatrix}, \quad \mathbf{X}^T\mathbf{X} = \begin{bmatrix} 12 & 4 & 4 & 6 \\ 4 & 4 & 0 & 2 \\ 4 & 0 & 4 & 2 \\ 6 & 2 & 2 & 6 \end{bmatrix}, \quad \mathbf{X}^T\mathbf{y} = \begin{bmatrix} 88.2 \\ 28.6 \\ 34.8 \\ 43.0 \end{bmatrix}$$

$$\text{hence} \quad \mathbf{b} = \begin{bmatrix} 6.383 \\ 0.950 \\ 2.500 \\ -0.367 \end{bmatrix}$$

and so $\mathbf{b}^T\mathbf{X}^T\mathbf{y} = 661.4133$.

(C) MODEL OMITTING EFFECTS OF LEVELS OF B:
$E(Y_{jkl}) = \mu + \alpha_j$ with $\alpha_1 = 0$

The design matrix is obtained by omitting the last three columns of matrix **X** in part (A) and so

$$\boldsymbol{\beta} = \begin{bmatrix} \mu \\ \alpha_2 \\ \alpha_3 \end{bmatrix}, \quad \mathbf{X}^{\mathrm{T}}\mathbf{X} = \begin{bmatrix} 12 & 4 & 4 \\ 4 & 4 & 0 \\ 4 & 0 & 4 \end{bmatrix}, \quad \mathbf{X}^{\mathrm{T}}\mathbf{y} = \begin{bmatrix} 88.2 \\ 28.6 \\ 34.8 \end{bmatrix}$$

$$\text{hence} \quad \mathbf{b} = \begin{bmatrix} 6.20 \\ 0.95 \\ 2.50 \end{bmatrix}$$

and $\mathbf{b}^{\mathrm{T}}\mathbf{X}^{\mathrm{T}}\mathbf{y} = 661.01$.

(D) MODEL OMITTING EFFECTS OF LEVELS OF A:
$E(Y_{jkl}) = \mu + \beta_k$ with $\beta_1 = 0$

The design matrix is given by columns 1 and 4 of matrix **X** in part (A) and so

$$\boldsymbol{\beta} = \begin{bmatrix} \mu \\ \beta_2 \end{bmatrix}, \quad \mathbf{X}^{\mathrm{T}}\mathbf{X} = \begin{bmatrix} 12 & 6 \\ 6 & 6 \end{bmatrix}, \quad \mathbf{X}^{\mathrm{T}}\mathbf{y} = \begin{bmatrix} 88.2 \\ 43.0 \end{bmatrix} \quad \text{hence} \quad \mathbf{b} = \begin{bmatrix} 7.533 \\ -0.367 \end{bmatrix}$$

and $\mathbf{b}^{\mathrm{T}}\mathbf{X}^{\mathrm{T}}\mathbf{y} = 648.6733$.

(E) MODEL WITH ONLY A MEAN EFFECT: $E(Y_{jkl}) = \mu$

In this case $\mathbf{b} = [\hat{\mu}] = 7.35$ and $\mathbf{b}^{\mathrm{T}}\mathbf{X}^{\mathrm{T}}\mathbf{y} = 648.27$.

A3.3 Orthogonal version obtained by a special choice of dummy variables

(A) FULL MODEL: $E(Y_{jkl}) = \mu + \alpha_j + \beta_k + (\alpha\beta)_{jk}$ with

$$\alpha_1 = \beta_1 = (\alpha\beta)_{11} = (\alpha\beta)_{12} = (\alpha\beta)_{21} = (\alpha\beta)_{31} = 0$$

$$\boldsymbol{\beta} = \begin{bmatrix} \mu \\ \alpha_2 \\ \alpha_3 \\ \beta_2 \\ (\alpha\beta)_{22} \\ (\alpha\beta)_{32} \end{bmatrix}, \quad \mathbf{X} = \begin{bmatrix} 1 & -1 & -1 & -1 & 1 & 1 \\ 1 & -1 & -1 & -1 & 1 & 1 \\ 1 & -1 & -1 & 1 & -1 & -1 \\ 1 & -1 & -1 & 1 & -1 & -1 \\ 1 & 1 & 0 & -1 & -1 & 0 \\ 1 & 1 & 0 & -1 & -1 & 0 \\ 1 & 1 & 0 & 1 & 1 & 0 \\ 1 & 1 & 0 & 1 & 1 & 0 \\ 1 & 0 & 1 & -1 & 0 & -1 \\ 1 & 0 & 1 & -1 & 0 & -1 \\ 1 & 0 & 1 & 1 & 0 & 1 \\ 1 & 0 & 1 & 1 & 0 & 1 \end{bmatrix};$$

The columns of matrix **X** corresponding to terms $(\alpha\beta)_{jk}$ are the products of columns corresponding to terms α_j and β_k.

$$\mathbf{X}^T\mathbf{X} = \begin{bmatrix} 12 & 0 & 0 & 0 & 0 & 0 \\ 0 & 8 & 4 & 0 & 0 & 0 \\ 0 & 4 & 8 & 0 & 0 & 0 \\ 0 & 0 & 0 & 12 & 0 & 0 \\ 0 & 0 & 0 & 0 & 8 & 4 \\ 0 & 0 & 0 & 0 & 4 & 8 \end{bmatrix}, \quad \mathbf{X}^T\mathbf{y} = \begin{bmatrix} 88.2 \\ 3.8 \\ 10.0 \\ -2.2 \\ 0.8 \\ 3.0 \end{bmatrix} \quad \text{hence} \quad \mathbf{b} = \begin{bmatrix} 7.35 \\ -0.2 \\ 1.35 \\ -0.1833 \\ -0.1167 \\ 0.4333 \end{bmatrix}$$

and so $\mathbf{b}^T\mathbf{X}^T\mathbf{y} = 662.62$.

(B) ADDITIVE MODEL: $E(Y_{jkl}) = \mu+\alpha_j+\beta_k$ with $\alpha_1 = \beta_1 = 0$

The design matrix is obtained by omitting the last two columns of matrix **X** in part (A). By the orthogonality of **X**, estimates of μ, α_2, α_3 and β_2 are the same as in part (A) and hence $\mathbf{b}^T\mathbf{X}^T\mathbf{y} = 661.4133$.

(C) MODEL OMITTING EFFECTS OF LEVELS OF B:
$E(Y_{jkl}) = \mu+\alpha_j$ with $\alpha_1 = 0$

The design matrix is obtained by omitting the last three columns of matrix **X** in part (A). By the orthogonality of **X**, estimates of μ, α_2 and α_3 are the same as in part (A) and $\mathbf{b}^T\mathbf{X}^T\mathbf{y} = 661.01$.

(D) MODEL OMITTING EFFECTS OF LEVELS OF A:
$E(Y_{jkl}) = \mu+\beta_k$ with $\beta_1 = 0$

As before the estimates of μ and β_2 are the same as in part (A) and $\mathbf{b}^T\mathbf{X}^T\mathbf{y} = 648.6733$.

(E) MODEL WITH ONLY A MEAN EFFECT: $E(Y_{jkl}) = \mu$

As before $\hat{\mu} = 7.35$ and $\mathbf{b}^T\mathbf{X}^T\mathbf{y} = 648.27$.

APPENDIX 4

Here are some log-linear models for three-dimensional contingency tables; this is not a complete list. The models are overparameterized so all the subscripted variables are subject to sum-to-zero or corner-point constraints.

A4.1 Three response variables

The multinomial distribution applies,

$$f(\mathbf{y};\boldsymbol{\theta}) = n! \prod_{j=1}^{J} \prod_{k=1}^{K} \prod_{l=1}^{L} \theta_{jkl}^{y_{jkl}} / y_{jkl}!.$$

(i) The maximal model is $E(Y_{jkl}) = n\,\theta_{jkl}$, i.e.

$$\eta_{jkl} = \mu+\alpha_j+\beta_k+\gamma_l+(\alpha\beta)_{jk}+(\alpha\gamma)_{jl}+(\beta\gamma)_{kl}+(\alpha\beta\gamma)_{jkl}$$

which has JKL independent parameters.

(ii) The *partial association* model is $E(Y_{jkl}) = n\theta_{jk.}\theta_{j.l}\theta_{.kl}$, i.e.

$$\eta_{jkl} = \mu+\alpha_j+\beta_k+\gamma_l+(\alpha\beta)_{jk}+(\alpha\gamma)_{jl}+(\beta\gamma)_{kl}$$

with $JKL-(J-1)(K-1)(L-1)$ independent parameters.

(iii) The *conditional independence* model in which, at each level of one variable, the other two are independent is, for example,

$$E(Y_{jkl}) = n\theta_{jk.}\theta_{j.l},$$

i.e.

$$\eta_{jkl} = \mu+\alpha_j+\beta_k+\gamma_l+(\alpha\beta)_{jk}+(\alpha\gamma)_{jl}$$

with $J(K+L-1)$ independent parameters.

(iv) A model with one variable independent of the other two, for example, $E(Y_{jkl}) = n\,\theta_{j..}\theta_{.kl}$, i.e.

$$\eta_{jkl} = \mu+\alpha_j+\beta_k+\gamma_l+(\beta\gamma)_{kl}$$

with $J+KL-1$ independent parameters.

(v) The *complete independence* model is $E(Y_{jkl}) = n\theta_{j..}\theta_{.k.}\theta_{..l}$, i.e.

$$\eta_{jkl} = \mu + \alpha_j + \beta_k + \gamma_l$$

with $J+K+L-2$ independent parameters.

(iv) *Non-comprehensive models* do not involve all variables, for example, $E(Y_{jkl}) = n\theta_{jk.}$ i.e.

$$\eta_{jkl} = \mu + \alpha_j + \beta_k + (\alpha\beta)_{jk}$$

with JK independent parameters.

A4.2 Two response variables and one explanatory variable

If the third variable is the fixed explanatory one, the product multinomial distribution is

$$f(\mathbf{y};\boldsymbol{\theta}) = \prod_{l=1}^{L} y_{..l}! \prod_{j=1}^{J}\prod_{k=1}^{K} \theta_{jkl}^{y_{jkl}}/y_{jkl}!$$

and all log-linear models must include the term $\mu + \gamma_l$.

(i) The maximal model is $E(Y_{jkl}) = y_{..l}\theta_{jkl}$, i.e.

$$\eta_{jk} = \mu + \alpha_j + \beta_k + \gamma_l + (\alpha\beta)_{jk} + (\alpha\gamma)_{jl} + (\beta\gamma)_{kl} + (\alpha\beta\gamma)_{jkl}$$

with JKL independent parameters.

(ii) The model describing independence of the response variables at each level of the explanatory variable is $E(Y_{jkl}) = y_{..l}\theta_{j.l}\theta_{.kl}$, i.e.

$$\eta_{jkl} = \mu + \alpha_j + \beta_k + \gamma_l + (\alpha\gamma)_{jl} + (\beta\gamma)_{kl}$$

with $L(J+K-1)$ independent parameters.

(iii) The *homogeneity* model in which the association between the responses is the same at each level of the explanatory variable is $E(Y_{jkl}) = y_{..l}\theta_{jk.}$, i.e.

$$\eta_{jkl} = \mu + \alpha_j + \beta_k + \gamma_l + (\alpha\beta)_{jk}$$

with $JK+L-1$ independent parameters.

A4.3 One response variable and two explanatory variables

If the first variable is the response the product multinomial distribution is

$$f(\mathbf{y};\boldsymbol{\theta}) = \prod_{k=1}^{K}\prod_{l=1}^{L} y_{.kl}! \prod_{j=1}^{J} \theta_{jkl}^{y_{jkl}}/y_{jkl}!$$

and all log-linear models must include the terms

$$\mu + \beta_k + \gamma_l + (\beta\gamma)_{kl}.$$

(i) The maximal model is $E(Y_{jkl}) = y_{.kl}\theta_{jkl}$, i.e.

$$\eta_{jkl} = \mu+\alpha_j+\beta_k+\gamma_l+(\alpha\beta)_{jk}+(\alpha\gamma)_{jl}+(\beta\gamma)_{kl}+(\alpha\beta\gamma)_{jkl}$$

with JKL independent parameters.

(ii) If the probability distribution is the same for all columns of each subtable then $E(Y_{jkl}) = y_{.kl}\theta_{j.l}$, i.e.

$$\eta_{jkl} = \mu+\alpha_j+\beta_k+\gamma_l+(\alpha\gamma)_{jl}+(\beta\gamma)_{kl}$$

with $L(J+K-1)$ independent parameters.

(iii) If the probability distribution is the same for all columns of every subtable then $E(Y_{jkl}) = y_{.kl}\theta_{j..}$, i.e.

$$\eta_{jkl} = \mu+\alpha_j+\beta_k+\gamma_l+(\beta\gamma)_{kl}$$

with $KL+J-1$ independent parameters.

REFERENCES

Andersen, E. B. (1980) *Discrete Statistical Models with Social Science Applications*, North Holland, Amsterdam.

Aranda-Ordaz, F. J. (1981) On two families of transformations to additivity for binary response data, *Biometrika*, **68**, 357–63.

Baker, R. J. and Nelder, J. A. (1978) *GLIM manual*, (release 3), Numerical Algorithms Group, Oxford.

Barndorff-Nielsen, O. (1978) *Information and Exponential Families in Statistical Theory*, Wiley, New York.

Belsley, D. A., Kuh, E. and Welsch, R. E. (1980) *Regression Diagnostics: Identifying Influential Data and Sources of Collinearity*, Wiley, New York.

Berkson, J. (1953) A statistically precise and relatively simple method of estimating the bio-assay with quantal response, based on the logistic function, *J. Amer. Statist. Assoc.*, **48**, 565–99.

Bliss, C. I. (1935) The calculation of the dosage–mortality curve, *Annals of Applied Biology*, **22**, 134–67.

Birch, M. W. (1963) Maximum likelihood in three-way contingency tables, *J.R. Statist. Soc. B*, **25**, 220–33.

Bishop, Y. M. M., Fienberg, S. E. and Holland, P. W. (1975) *Discrete Multivariate Analysis: Theory and Practice*, M.I.T. Press, Cambridge, Mass.

Charnes, A., Frome, E. L. and Yu, P. L. (1976) The equivalence of generalized least squares and maximum likelihood estimates in the exponential family, *J. Amer. Statist. Assoc.* **71**, 169–71.

Copenhaver, T. W. and Mielke, P. W. (1977) Quantit analysis: a quantal assay refinement, *Biometrics*, **33**, 175–86.

Chambers, J. M. (1973) Fitting non-linear models: numerical techniques, *Biometrika*, **60**, 1–13.

Cox, D. R. (1966) A simple example of a comparison involving quantal data, *Biometrika*, **53**, 215–20.

Cox, D. R. (1970) *The Analysis of Binary Data*, Chapman and Hall, London.

Cox, D. R. and Hinkley, D. V. (1974) *Theoretical Statistics*, Chapman and Hall, London.

Cox, D. R. and Snell, E. J. (1968) A general definition of residuals, *J.R. Statist. Soc. B*, **30**, 248–75.

Draper, N. R. and Smith, H. (1981) *Applied Regression Analysis*, 2nd edn., Wiley, New York.

Everitt, B. S. (1977) *The Analysis of Contingency Tables*, Chapman and Hall, London.

Fienberg, S. E. (1977) *The Analysis of Cross-Classified Categorical Data*, M.I.T. Press, Cambridge, Mass.

Finney, D. J. (1973) *Statistical Method in Biological Assay*, 2nd edn., Hafner, New York.

Gillis, P. R. and Ratkowsky, D. A. (1978) The behaviour of estimators of parameters of various yield–density relationships, *Biometrics*, **34**, 191–8.

Gordon, T. and Foss, B. M. (1966) The role of stimulation in the delay of onset of crying in the new-born infant, *J. Expt. Psychol.* **16**, 79–81.

Graybill, F. A. (1976) *Theory and Application of the Linear Model*, Duxbury, N. Scituate, Mass.

Grizzle, J. E., Starmer, C. F. and Koch, G. G. (1969) Analysis of categorical data by linear models, *Biometrics*, **25**, 489–504.

Holliday, R. (1960) Plant population and crop yield, *Field Crop Abstracts*, **13**, 159–67, 247–54.

Larntz, K. (1978) Small sample comparisons of exact levels for chi-square goodness-of-fit statistics, *J. Amer. Statist. Assoc.*, **73**, 253–63.

McCullagh, P. (1980) Regression models for ordinal data, *J.R. Statist. Soc., B*, **42**, 109–42.

McCullagh, P. and Nelder, J. A. (1983) *Generalized Linear Models*, Chapman and Hall, London.

McKinlay, S. M. (1978) The effect of nonzero second-order interaction on combined estimators of the odds ratio, *Biometrika*, **65**, 191–202.

Nelder, J. A. (1974) Log linear models for contingency tables: a generalization of classical least squares, *Appl. Statist.*, **23**, 323–9.

Nelder, J. A. and Wedderburn, R. W. M. (1972) Generalised linear models, *J.R. Statist. Soc. A*, **135**, 370–84.

Neter, J. and Wasserman, W. (1974) *Applied Linear Statistical Models*, Irwin, Homewood, Illinois.

Pregibon, D. (1980) Goodness of link tests for generalised linear models, *Appl. Statist.*, **29**, 15–24.

Prentice, R. L. (1976) A generalisation of the probit and logit methods for dose response curves, *Biometrics*, **32**, 761–8.

Ratkowsky, D. A. and Dolby, G. R. (1975) Taylor series linearization and scoring for parameters in nonlinear regression, *Appl. Statist.*, **24**, 109–11.

Roberts, G., Martyn, A. L., Dobson, A. J. and McCarthy, W. H. (1981) Tumour thickness and histological type in malignant melanoma in New South Wales, Australia, 1970–76, *Pathology*, **13**, 763–70.

Sangwan-Norrell, B. S. (1977) Androgenic stimulating factors in the anther and isolated pollen grain culture of *Datura innoxia* Mill., *J. Expt. Botany*, **28**, 843–52.

Scheffé, H. (1959) *The Analysis of Variance*, Wiley, New York.

Searle, S. R. (1971) *Linear Models*, Wiley, New York.

Seber, G. A. F. (1977) *Linear Regression Analysis*, Wiley, New York.

Whittaker, J. and Aitkin, M. (1978) A flexible strategy for fitting complex log-linear models, *Biometrics*, **34**, 487–95.

Winer, B. J. (1971) *Statistical Principles in Experimental Design*, 2nd edn., McGraw-Hill, New York.

Wood, C. L. (1978) Comparison of linear trends in binomial proportions, *Biometrics*, **34**, 496–504.

Woolf, B. (1955) On estimating the relation between blood group and disease, *Ann. Human Genet., Lond.*, **19**, 251–3.

INDEX